Proceedings Symposium on Value Distribution Theory in Several Complex Variables

Notre Dame Mathematical Lectures
Number 12

Proceedings Symposium on Value Distribution Theory in Several Complex Variables

on the occasion of the inauguration of Wilhelm Stoll as the Vincent F. Duncan and Annamarie Micus Duncan Professor of Mathematics

April 28/29, 1990
University of Notre Dame

Edited by
Wilhelm Stoll

UNIVERSITY OF NOTRE DAME PRESS
NOTRE DAME, INDIANA

Library of Congress Cataloging-in-Publication Data

Symposium on Value Distribution Theory in Several Complex Variables (1990 : University of Notre Dame)

Proceedings : Symposium on Value Distribution Theory in Several Complex Variables : on the occasion of the inauguration of Wilhelm Stoll as the Vincent F. Duncan and Annamarie Micus Duncan professor of mathematics. April 28/29 1990, University of Notre Dame.

p. cm. — (Notre Dame mathematical lectures ; no. 12)

ISBN 0-268-01512-0

1. Value distribution theory—Congresses. 2. Functions of several complex variables—Congresses. I. Stoll, Wilhelm, 1923–
II. Series
QA1.N87 no. 12
[QA331.7]
510 s—dc20 91-42751
[515".94] CIP

Manufactured in the United States of America

CONTENTS

FOREWORD

In 1988 Vincent J. Duncan and Annamarie Micus Duncan sponsored a chair at the University of Notre Dame in honor of Walter Duncan, a 1912 graduate of Notre Dame and a long time trustee of the University. In the fall of that year I was appointed to this chair. In conjunction with the inaugural lecture, the University of Notre Dame held a symposium on value distribution in several complex variables April 28/29, 1990. It was to reflect the growth of this field from its beginning about 60 years ago as well as its connections to related areas. The Symposium was solely funded by Notre Dame. Thus only speakers present within the USA at the time could be invited and support was restricted to them. Professor Shiing-shen Chern, who contributed substantially to the field and helped to build up Mathematics at Notre Dame, declined to lecture, but he participated actively in the Symposium as an honored guest. Professor Phillip Griffiths made significant contributions to the field, but in the end, his obligation as Provost of Duke University prevented him from coming. The names of the other invited speakers and the title of their talks can be found in the Symposium program reprinted below. Two significant lectures preceded the Symposium the day before. Sponsored by the Physics Department, Paul Chu gave an University wide address on superconductivity. Professor Chern attended his son in law's lecture. Dr. Peter Polyakov, a close collaborator of Professor Henkin and a recent immigrant to the USA, spoke about a proof of a 1973 conjecture of mine in a mathematics colloquium. The Symposium was well attended, but, unfortunately no precise list of participants was kept.

In these proceedings, some speakers wrote on the topics of their lectures, some others selected different topics and some did not send contributions. Serge Lang, who had spoken on William Cherry's results, requested, that in the Proceedings these results should be communicated directly. My inaugural address in the Proceedings is an expanded version of the original one.

I thank all those who made this Symposium and these Proceedings possible and those who contributed and helped with it. In particular I thank the donors for their generosity.

Wilhelm Stoll
July 1991

PROGRAM

Saturday, April 28, 1990

9:00–9:15 am	Introducing Remarks Timothy O'Meara, Provost, University of Notre Dame Francis Castellino, Dean, College of Science
9:15–10:15 am	Inaugural Lecture Wilhelm Stoll, Notre Dame Title: High Points in the History of Value Distribution Theory of Several Complex Variables
10:30–11:30 am	Bernard Shiffman, Johns Hopkins Title: Bounds On the Distance to Algebraic Varieties in $\mathbb{C}^n$
12:00 noon	Lunch - Morris Inn
2:00–3:00 pm	Yum Tong Siu, Harvard Title: Some Recent Results On Non-equidimensional Value Distribution Theory
3:15–4:15 pm	David Drasin, Purdue Title: The Branching Term of Nevanlinna's Theory
4:30–5:30 pm	Paul Vojta, Institute for Advanced Study Title: Recent Work On Nevanlinna Theory and Diophantine Approximations
6:30 pm	Dinner - Morris Inn

Sunday, April 29, 1990

9:15–10:15 am	Serge Lang, Yale Title: The Error Term in Nevanlinna Theory
10:30–11:30 am	Pit-Mann Wong, Notre Dame Title: Second Main Theorems of Nevanlinna's Theory
12:00 noon	Lunch - Morris Inn
1:45–2:45 pm	Lo Yang, Notre Dame/Academia Sinica, Beijing Title: Some Results and Problems in the Theory of Value Distribution
3:00–4:00 pm	Gennadi Henkin, Notre Dame/Academy of Science USSR Title: The Characterization of the Scattering Datas for the Schrödinger Operator in Terms of the $\bar{\partial}$ - Equation and Growth Conditions

HIGH POINTS IN THE HISTORY OF VALUE DISTRIBUTION THEORY OF SEVERAL COMPLEX VARIABLES

Wilhelm Stoll

Inaugural Lecture

Timothy O'Meara and Frank Castellino thank your for your kind introduction. I am deeply moved by your words and by the appointment to the chair. Foremost I thank the donors Vincent J. Duncan and Annamarie Micus Duncan for their generosity. My colleagues and I are most grateful for this recognition of our work by the donors and the administration of the University.

Ladies and gentlemen, colleagues, speakers and participants! This inaugural address opens the *Symposium on Value Distribution Theory in Several Complex Variables* sponsored by the University of Notre Dame. Welcome to all of you. An inaugural address, an Antrittsvorlesung, so late in life seems to be out of place and perhaps should be called an Abschiedsvorlesung. Yet, hopefully, this is premature and I can be around a few more years. Taking the hint, I will look backwards and recall some of the high points in the development of the theory. Time permits only a few topics.

Looking backwards, out of the mist of time there emerges not an abstract theory but the lively memory of those who taught me mathematics: Siegfried Kerridge, Wilhelm Germann, Wilhelm Schweizer and later at the University Hellmuth Kneser, Konrad Knopp, Erich Kamke, G. G. Lorentz and Max Müller. Also there appear those who inspired me but who were not directly my teachers: Heinz Hopf, Hermann Weyl, Rolf Nevanlinna and one who is right here with us: Shiing-shen Chern, we all welcome you. Thirty years ago you recruited me for Notre Dame. You supported the growth of this department in many ways. Your work on value distribution in several

---

This research was supported in part by the National Science Foundation Grant DMS-87-02144.

complex variables counts as one of your many marvelous contributions to mathematics. Thank you for coming.

The giants of the 19th century created the theory of entire functions. In this century, in 1925, with a stroke of genius, Rolf Nevanlinna extended this theory to a value distribution theory of meromorphic functions. His two *Main Theorems* are the foundation upon which Nevanlinna theory rests.

In 1933, Henri Cartan [8] proved Nevanlinna's Second Main Theorem for the case of holomorphic curves. If we view curves belonging to the theory of several dependent variables, then Cartan's paper provides the first theorem in the theory of value distribution in several complex variables. Thus let me outline his result. However, I shall use today's terminology and advancement.

For each $0 < r \in \mathbb{R}$ define the discs and circle

$$\text{(1)} \quad \mathbb{C}[r] = \{z \in \mathbb{C} \mid |z| \leq r\} \quad \mathbb{C}(r) = \{z \in \mathbb{C} \mid |z| < r\}$$

$$\text{(2)} \quad \mathbb{C}\langle r\rangle = \{z \in \mathbb{C} \mid |z| = r\} \qquad \mathbb{C}_* = \mathbb{C} - \{0\}.$$

An integral valued function $\nu : \mathbb{C} \to \mathbb{Z}$ is said to be a *divisor* if

$$\text{(3)} \qquad S = \operatorname{supp}\nu = \operatorname{clos}\{z \in \mathbb{C} \mid \nu(z) \neq 0\}$$

is a closed set of isolated points in $\mathbb{C}$. For all $r \geq 0$ the *counting function* n_ν of ν is defined by the finite sum

$$\text{(4)} \qquad n_\nu(r) = \sum_{z \in \mathbb{C}[r]} \nu(z).$$

For $0 < s < r \in \mathbb{R}$, the *valence function* N_ν of ν is defined by

$$\text{(5)} \qquad N_\nu(r, s) = \int_s^r n_\nu(t) \frac{dt}{t}.$$

If $h \not\equiv 0$ is an entire function, let $\mu_h(z)$ be the *zero-multiplicity* of h at z. Then $\mu_h : \mathbb{C} \to \mathbb{Z}$ is a non-negative divisor called the *zero divisor* of h.

The exterior derivative $d = \partial + \bar{\partial}$ on differential forms twists to

$$d^c = \frac{i}{4\pi}(\bar{\partial} - \partial) \tag{6}$$

on complex manifolds. Define $\tau_o : \mathbb{C} \to \mathbb{R}$ by $\tau_o(z) = |z|^2$ for $z \in \mathbb{C}$. Define

$$\sigma = d^c \log \tau_0, \tag{7}$$

If $r > 0$, then

$$\int_{\mathbb{C}\langle r\rangle} \sigma = 1. \tag{8}$$

If $h \not\equiv 0$ is an entire function and if $r > o$, the **Jensen Formula**

$$N_{\mu_h}(r, s) = \int_{\mathbb{C}\langle r\rangle} \log h\sigma - \int_{\mathbb{C}\langle s\rangle} \log h\sigma \tag{9}$$

is a forerunner of Nevanlinna's First Main Theorem.

Let V be a normed, complex vector space of finite dimension $n + 1 > 1$. Put $V_* = V - \{0\}$. Then the multiplicative group $\mathbb{C}_*$ acts on V_*. The quotient space $\mathbb{P}(V) = V_*/\mathbb{C}_*$ is the associated projective space. The quotient map $\mathbb{P} : V_* \to \mathbb{P}(V)$ is open and holomorphic. If $M \subseteq V$, put $\mathbb{P}(M) = P(M \cap V_*)$. If W is a linear subspace of V with dimension $p + 1$, then $\mathbb{P}(W)$ is called a p-*plane* of $\mathbb{P}(V)$. If $p = n - 1$, then $\mathbb{P}(W)$ is called a *hyperplane*. The *dual* complex vector space V^* of V consists of all $\mathbb{C}$-linear functions $\mathfrak{a} : V \to \mathbb{C}$. Here $\|\mathfrak{a}\|$ is the smallest real number such that $|\mathfrak{a}(\mathfrak{x})| \le \|\mathfrak{a}\| \; \|\mathfrak{x}\|$ for all $\mathfrak{x} \in V$. Then $\| \; \|$ is a norm on V^*. Also write $\langle\mathfrak{x}, \mathfrak{a}\rangle = \mathfrak{a}(\mathfrak{x})$. Here $\langle\mathfrak{x}, \mathfrak{a}\rangle = \langle\mathfrak{a}, \mathfrak{x}\rangle$ indicates $(V^*)^* = V$. If $a = \mathbb{P}(\mathfrak{a}) \in \mathbb{P}(V^*)$, then $E[a] = \mathbb{P}(\ker \mathfrak{a})$ is a hyperplane in $\mathbb{P}(V)$. The assignment $a \to E[a]$ parameterizes the set of hyperplanes bijectively. The *distance* from $x = \mathbb{P}(\mathfrak{x}) \in \mathbb{P}(V)$ to $E[a]$ is measured by

$$0 \le \Box x, a\Box = \frac{|\langle\mathfrak{x}, \mathfrak{a}\rangle|}{\|\mathfrak{x}\| \; \|\mathfrak{a}\|} \le 1. \tag{10}$$

Let $f : \mathbb{C} \to \mathbb{P}(V)$ be a holomorphic map. A holomorphic map $\mathfrak{v} : \mathbb{C} \to V_*$ is called a *reduced representation* of f if $\mathbb{P} \circ \mathfrak{v} = f$. A reduced representation exists. Then $\mathfrak{w} : \mathbb{C} \to V_*$ is a reduced representation of f if and only if there is a holomorphic function $h : \mathbb{C} \to \mathbb{C}_*$ without zeroes such that $\mathfrak{w} = h\mathfrak{v}$. For $0 < s < r \in \mathbb{R}$ the *characteristic function* of f is defined by

$$(11) \qquad T_f(r, s) = \int_{\mathbb{C}<r>} \log \|\mathfrak{v}\| \sigma - \int_{\mathbb{C}<s>} \log \|\mathfrak{v}\| \sigma.$$

By (9), $T_f(r, s)$ does not depend on the choice of $\mathfrak{v}$, Since $\log \|\mathfrak{v}\|$ is subharmonic, $T_f \geq 0$. If f is constant, $\mathfrak{v}$ can be taken as a constant. Hence $T_f(r, s) \equiv 0$. If f is not constant, then $T_f(r, s) > 0$ and $T_f(r, s) \to \infty$ for $r \to \infty$. If $\| \ \|$ and $|\!|\!| \ |\!|\!|$ are two norms on V, there are constants $C_2 \geq C_1 > 0$ such that $C_1 |\!|\!|\mathfrak{x}|\!|\!| \leq \|\mathfrak{x}\| \leq C_2 |\!|\!|\mathfrak{x}|\!|\!|$ for all $\mathfrak{x} \in V$. Put $C = \log C_2/C_1 \geq 0$. If $0 < s < r$, then

$$(12) \qquad |T_f(r, s, \| \ \|) - T_f(r, s, |\!|\!| \ |\!|\!|)| \leq C.$$

Let $f : \mathbb{C} \to \mathbb{P}(V)$ and $g : \mathbb{C} \to \mathbb{P}(V^*)$ be holomorphic maps. They are called *free* if $f(z) \notin E[g(z)]$ for some $z \in \mathbb{C}$. Take reduced representations $\mathfrak{v}$ of f and $\mathfrak{w}$ of g, then (f, g) is free if and only if $<\mathfrak{v}, \mathfrak{w}> = h \not\equiv 0$. If so, the *intersection divisor* $\mu_{f,g} = \mu_h \geq 0$ does not depend on the choices of $\mathfrak{v}$ and $\mathfrak{w}$. Its *counting function* and its *valence function* are abbreviated by $n_{f,g}$ and $N_{f,g}$ respectively. The pair (f, g) is free if and only if $\Box f, g \Box \not\equiv 0$. If so, for $r > 0$ the *compensation function* $m_{f,g}$ of (f, g) is defined by

$$(13) \qquad m_{f,g}(r) = \int_{\mathbb{C}<r>} \log \frac{1}{\Box f, g \Box} \sigma \geq 0.$$

For $0 < s < r$, the identities (9), (11) and (13) imply the **First Main Theorem**

$$(14) \qquad T_f(r, s) + T_g(r, s) = N_{f,g}(r, s) + m_{f,g}(r) - m_{f,g}(s).$$

Cartan [8] considered the case of constant $g \equiv a \in \mathbb{P}(V^*)$ only which yields

(15) $$T_f(r,s) = N_{f,a}(r,s) + m_{f,a}(r) - m_{f,a}(s)$$

which Cartan [8] mentions only implicitely. If $n = 1$, Rolf Nevanlinna proved (15) in [32] (1925).

If f or g or both are not constant and if (f, g) is free the *defect* is defined by

(16) $$\begin{aligned} 0 \leq \delta(f,g) &= \underline{\lim}_{r\to\infty} \frac{m_{f,g}(r)}{T_f(r,s)+T_g(r,s)} \\ &= 1 - \overline{\lim}_{r\to\infty} \frac{N_{f,g}(r,s)}{T_f(r,s)+T_g(r,s)} \leq 1 \end{aligned}$$

The map g is said to *grow slower* than f, if $T_g(r,s)/T_f(r,s) \to 0$ for $r \to \infty$. By (12), the defect does not depend on the choice of the norm on V. Also the defect is independent of s. Observe that $\mu_{f,g} = \mu_{g,f}, n_{f,g} = n_{g,f}, N_{f,g} = N_{g,f}, m_{f,g} = m_{g,f}$ and $\delta(f,g) = \delta(g,f)$. Since most investigators concentrate on constant g or on the case where g grows slower than f, this symmetry is little known.

Since the choice of the norm on V does not matter, we can choose a hermitian norm which comes from a positive definite hermitian form $(\cdot|\cdot) : V \times V \to \mathbb{C}$ with $\|\mathfrak{x}\|^2 = (\mathfrak{x}|\mathfrak{x})$ for $\mathfrak{x} \in V$. Define $\tau : V \to \mathbb{C}$ by $\tau(\mathfrak{x}) = \|\mathfrak{x}\|^2$ for $\mathfrak{x} \in V$. Then τ is of class C^∞. There is one and only one positive form Ω of bidegree(1,1) on $\mathbb{P}(V)$, called the *Fubini Study form* such that $dd^c \log \tau = \mathbb{P}^*(\Omega)$ on V_*. Let $\mathfrak{v} : \mathbb{C} \to V_*$ be a reduced representation of f. Then $f = \mathbb{P} \circ \mathfrak{v}$ implies

(17) $$dd^c \log \|\mathfrak{v}\|^2 = \mathfrak{v}^*(\mathbb{P}^*(\Omega)) = f^*(\Omega).$$

If Stokes theorem and fiber integration are applied to (11) we obtain the *Ahlfors-Shimizu* definition of the *characteristic function* of f

(18) $$T_f(r,s) = \int_s^r \int_{\mathbb{C}[t]} f^*(\Omega) \frac{dt}{t} \quad \text{for } 0 < s < r.$$

Here $A_f(t) = \int_{\mathbb{C}[t]} f^*(\Omega) \geq 0$ increases. Put $A_f(\infty) = \lim_{t\to\infty} A_f(t) \leq \infty$. Then

(19) $$\lim_{r\to\infty} \frac{T_f(r,s)}{\log r} = A_f(\infty).$$

Now f is constant if and only if $A_f(\infty) = 0$ and f is rational if and only if $A_f(\infty) < \infty$.

Let $\mathfrak{A} = \{a_j\}_{j\in Q}$ be a family of points $a_j \in \mathbb{P}(V^*)$ representating hyperplanes. If $P \subseteq Q$, define $\mathfrak{A}_P = \{a_j\}_{j\in P}$. For each $j \in Q$ pick $\mathfrak{a}_j \in V_*^*$ with $a_j = \mathbb{P}(\mathfrak{a}_j)$. Our definitions will not depend on the choice of $\mathfrak{a}_j$. Put $q = \#Q$. Then $\mathfrak{A}$ is said to be *linearly independent* if there is a bijective map $\lambda : \mathbb{N}[1,q] \to Q$ such that $\mathfrak{a}_{\lambda(1)}, \dots, \mathfrak{a}_{\lambda(q)}$ are linearly independent. If so, then $q \leq n+1$. Moreover $\mathfrak{A}$ is said to be *basic* if $\mathfrak{A}$ is linearly independent and $q = n+1$. Moreover $\mathfrak{A}$ is said to be in *general position* if $\mathfrak{A}_P$ is linearly independent for each $P \subseteq Q$ with $0 < \#P \leq n+1$. If N is an integer and if $q > N \geq n$, then $\mathfrak{A}$ is said to be in *N-subgeneral position* (Chen [9]) if for every subset S of Q with $\#S = N+1$, there is a subset P of S such that $\mathfrak{A}_P$ is basic.

Let $f : \mathbb{C} \to \mathbb{P}(V)$ be a holomorphic map. Then there is a unique linear subspace W of smallest dimension $k+1$ of V such that $f(\mathbb{C}) \subseteq \mathbb{P}(W)$. Then f is said to be *k-flat*. If $k = n$, then $W = V$ and f is said to be *linearly non-degenerated.*

Take $0 \leq s \in \mathbb{R}$. Let $G : \mathbb{R}[s,+\infty) \to \mathbb{R}$ and $H : \mathbb{R}[s,+\infty)$ be functions. Then $G \underset{\cdot\cdot}{\leq} H$ means that there is a subset E of finite measure of $\mathbb{R}_+ = \mathbb{R}[0,+\infty)$ such that $G(r) \leq H(r)$ for all $r \in \mathbb{R}[s,+\infty) - E$.

Second Main Theorem (Cartan [8] 1933)

Let V be a hermitian vector space of dimension $n+1 > 1$. Let $f : \mathbb{C} \to \mathbb{P}(V)$ be a linearly non-degenerated, holomorphic map. Let $\mathfrak{A} = \{a_j\}_{j\in Q}$ be a finite family of "hyperplanes" $a_j \in \mathbb{P}(V^*)$ in general position with $n+1 < q = \#Q < \infty$. Take $s > 0$ and $\varepsilon > 0$. Then

(20) $$\sum_{j\in Q} m_{f,a_j}(r) \underset{\cdot\cdot}{\leq} (n+1)T_f(r,s) + \frac{1}{2}n(n+1)(1+\varepsilon)\log T_f(r,s) + \varepsilon \log r.$$

As a consequence, we obtain trivially

Defect Relation (Cartan [8] 1933)

Under the assumptions of the Second Main Theorem we have

$$\sum_{j\in Q}\delta(f,a_j)\leq n+1. \tag{21}$$

If $f:\mathbb{C}\to\mathbb{P}(V)$ is only k-flat, and if $\mathfrak{A}$ is in general position such that (f,a_j) is free for each $j\in Q$, Henri Cartan conjectured in 1933 that

$$\sum_{j\in Q}\delta(f,a_j)\leq 2n-k+1, \tag{22}$$

which was proven by Nochka [35] in 1982. Thus if $\#Q\geq 2n+1$ and $f(\mathbb{C})\cap E[a_j]=0$ for all $j\in Q$, then $2n+1\leq 2n-k+1$. Therefore $k=0$ and f is constant. Hence

$$\mathbb{P}(V)-\bigcup_{j\in Q}E[a_j] \tag{23}$$

is Brody-hyperbolic. In fact by a theorem of Chen [9] (22) can be improved:

Defect Relation of Cartan-Nochka-Chen

Let V be a hermitian vector space of dimension $n+1>1$. Let $f:\mathbb{C}\to\mathbb{P}(V)$ be a k-flat, holomorphic map. Let $\mathfrak{A}=\{a_j\}_{j\in Q}$ be a finite family of "hyperplanes" $a_j\in\mathbb{P}(V^*)$ in N-subgeneral position with $N\geq n$ and $N+1\leq\#Q=q<\infty$. Assume that (f,a_j) is free for each $j\in Q$. Then

$$\sum_{j\in Q}\delta(f,a_j)\leq 2N-k+1. \tag{24}$$

An alternative proof of the defect relation (21) was given by Ahlfors [1] in 1941. Also he proves a defect relation for associated maps. His proof is very powerful and works in more general situations.

Hermann and Joachim Weyl [90] lifted Ahlfors's proof to Riemann surfaces. It was simplified by H. Wu [92] in 1970, Cowen and Griffiths [17] in 1976 and Pit-Mann Wong [93] in 1976. I extended this Ahlfors-Weyl theory to non-compact Kaehler manifolds [65]. However first we have to inquire how value distribution was extended to functions and maps of several independent complex variables.

Hellmuth Kneser created such an extension in two fundamental papers [23] in 1936 and [24] in 1938. Although these papers are little remembered today, they still influence the present research in value distribution of several independent complex variables. Therefore let me explain his fundamental ideas. Again I will cast them in modern terminology and perspective.

Let M be a connected, complex manifold of dimension m. Let $f \not\equiv 0$ be a holomorphic function on M. Take $p \in M$. Let $\alpha : U' \to U$ be a biholomorphic map of an open ball U' in $\mathbb{C}^m$ centered at 0 onto an open subset U of M with $\alpha(0) = p$. Then for each integer $\lambda \geq 0$ there is a unique homogeneous polynomial P_λ of degree λ such that

$$f \circ \alpha = \sum_{\lambda=0}^{\infty} P_\lambda \tag{25}$$

where the convergence is uniform on every compact subset of U'. Since $f|U \not\equiv 0$, there is a unique number $\mu = \mu_f(p) \geq 0$ depending on f and p only such that $P_\mu \not\equiv 0$ and $P_\lambda \equiv 0$ for all $\lambda \in \mathbb{Z}$ with $0 \leq \lambda < \mu$. The number $\mu_f(p)$ is called the *zero-multiplicity* of f at p and the function $\mu_f : M \to \mathbb{Z}$ is called the *zero-divisor* of f.

An integral valued function $\nu : M \to \mathbb{Z}$ is said to be a *divisor* on M if and only if for every point $p \in M$ there is an open, connected neighborhood U of p with holomorphic functions $g \not\equiv 0$ and $h \not\equiv 0$ on U such that

$$\nu|U = \mu_g - \mu_h. \tag{26}$$

Let S be the support of ν. Then $S = \emptyset$ if and only if $\nu \equiv 0$. If $S \neq \emptyset$, then S is a pure (m$-$1)-dimensional analytic subset of M. Let $\mathfrak{R}(S)$ be the set of *regular points* of S and let $\sum(S) = S - \mathfrak{R}(S)$ be the set of *singular points* of S. Then $\nu|\mathfrak{R}(S)$ is locally constant.

Let $\tau : M \to \mathbb{R}_+$ be an unbounded, non-negative function of class C^∞ on M. If $B \subseteq M$ and $0 \leq r \in \mathbb{R}$, abbreviate

$$B[r] = \{x \in B | \tau(x) \leq r^2\} \quad B(r) = \{x \in B | \tau(x) < r^2\} \tag{27}$$

$$B\langle r\rangle = \{x \in B | \tau(x) = r^2\} \quad B_* = \{x \in B | \tau(x) > 0\} \tag{28}$$

Here τ is called an *exhaustion* of M if and only if $M[r]$ is compact for each $r > 0$. Abbreviate

$$\upsilon = dd^c\tau \quad \omega = dd^c \log \tau \quad \sigma = d^c \log \tau \wedge \omega^{m-1} \tag{29}$$

Then $d\sigma = \omega^m$. The function τ is said to be *parabolic* if and only if

$$\omega \geq 0 \quad \omega^m \equiv 0 \quad \upsilon^m \not\equiv 0. \tag{30}$$

If so, then $\upsilon \geq 0$. More over τ is said to be *strictly parabolic* if and only if τ is parabolic and $\upsilon > 0$ on M. If τ is an exhaustion and parabolic, then (M, τ) is said to be a *parabolic manifold*. If so, there is a constant $\varsigma > 0$ such that

$$\int_{M[r]} \upsilon^m = \varsigma r^{2m} \tag{31}$$

for all $r \geq 0$. Then for almost all $r > 0$ we have

$$\int_{M\langle r\rangle} \sigma = \varsigma. \tag{32}$$

In 1973 Griffiths and King [19] introduced parabolic manifolds. The concept was expanded in [75]. If τ is an exhaustion and strictly parabolic function, (M, τ) is said to be *a strictly parabolic manifold.* In [77] 1980 I showed that (M, τ) is strictly parabolic if and only if there is a hermitian vector space W of dimension m and a biholomorphic map $h : M \to W$ such that $\tau = \|h\|^2$. We assume that (M, τ) is strictly parabolic and we identify $M = W$ such that h becomes the identity. In this case $\varsigma = 1$ and M is a hermitian vector space, which was Kneser's starting point. We assume that $m > 1$. If $u : M\langle 1\rangle \to \mathbb{C}$ is a function such that $u\sigma$ is integrable over the unit sphere $M\langle 1\rangle$, the *mean value* of u is defined by

$$\mathfrak{M}(u) = \int_{M\langle 1\rangle} u\sigma \tag{33}$$

Let V be a hermitian vector space of dimension $n + 1 > 1$. Let $f : M \to \mathbb{P}(V)$ and $g : M \to \mathbb{P}(V^*)$ be meromorphic maps. Let I_f and I_g be the indeterminacies of f and g respectively. Then (f, g) is said to be *free* if there exists $z \in M - I_f \cup I_g$ such that $f(z) \notin E[g(z)]$. For each "unit" vector $\mathfrak{b} \in M\langle 1\rangle$ an isometric embedding $j_\mathfrak{b} : \mathbb{C} \to M$ is defined by $j_\mathfrak{b}(z) = z\mathfrak{b}$ for $z \in \mathbb{C}$. If $j_\mathfrak{b}(\mathbb{C}) \not\subseteq I_f \cup I_g$ the pull back holomorphic maps $f_\mathfrak{b} = j_\mathfrak{b}^*(f) : \mathbb{C} \to \mathbb{P}(V)$ and $g_\mathfrak{b} = j_\mathfrak{b}^*(g) : \mathbb{C} \to \mathbb{P}(V^*)$ exist and $(f_\mathfrak{b}, g_\mathfrak{b})$ is free for almost all $\mathfrak{b} \in M\langle 1\rangle$. If $0 < s < r$ the **First Main Theorem** holds

$$T_{f_\mathfrak{b}}(r, s) + T_{g_\mathfrak{b}}(r, s) = N_{f_\mathfrak{b}, g_\mathfrak{b}}(r, s) + m_{f_\mathfrak{b}, g_\mathfrak{b}}(r) - m_{f_\mathfrak{b}, g_\mathfrak{b}}(s) \tag{34}$$

Now Kneser [24] applied the operator $\mathfrak{M}$ termwise in (34) to obtain the respective value distribution functions and the **First Main Theorem**

$$T_f(r, s) + T_g(r, s) = N_{f,g}(r, s) + m_{f,g}(r) - m_{f,g}(s) \tag{35}$$

Of course Kneser considered the case $n = 1$ only. Then f is a meromorphic function. Also he assumed that $g \equiv a \in \mathbb{P}_1$ is constant. Had he stopped with the above derivation of (35), his result would have been worthless. He proceeded and expressed the value distribution functions in meaningful analytic and geometric terms. This made the paper successful.

Let Ω be the Fubini Study form on $\mathbb{P}(V)$. For $t > 0$ define A_f by

$$A_f(t) = \frac{1}{t^{2m-2}} \int_{M[t]} f^*(\Omega) \wedge \upsilon^{m-1} \geq 0. \tag{36}$$

He showed that A_f increases. Hence the limits

$$0 \le \lim_{t\to 0} A_f(t) = A_f(0) < \infty \tag{37}$$
$$0 \le \lim_{t\to\infty} A_f(t) = A_f(\infty) \le \infty$$

exist. Kneser obtained the identity

$$A_f(t) = \int_{M[t]} f^*(\Omega) \wedge \omega^{m-1} + A_f(0) \tag{38}$$

for $t > 0$. Here f is constant if and only if $A_f(\infty) = 0$ and f is rational if and only if $A_f(\infty) < \infty$. Kneser proved

$$T_f(r, s) = \int_s^r A_f(t)\frac{dt}{t} \tag{39}$$

for $0 < s < r$. Moreover we have

$$\lim_{r\to\infty} \frac{T_f(r, s)}{\log r} = A_f(\infty). \tag{40}$$

A holomorphic map $\mathfrak{v} : M \to V$ is said to be a *reduced representation* of f if and only if $\dim \mathfrak{v}^{-1}(0) \le m-2$ and $f(z) = \mathbb{P}(\mathfrak{v}(z))$ for all $z \in M - I_f$ with $\mathfrak{v}(z) \ne 0$. In fact $I_f = \mathfrak{v}^{-1}(0)$. Reduced representations exist since M is a vector space. If $\mathfrak{v}$ is a reduced representation of f, any other reduced representation is given by $h\mathfrak{v}$, where $h : M \to \mathbb{C}_*$ is an entire function without zeroes. If $0 < s < r$, then

$$T_f(r, s) = \int_{M\langle r\rangle} \log \|\mathfrak{v}\|\sigma - \int_{M\langle s\rangle} \log \|\mathfrak{v}\|\sigma. \tag{41}$$

Since (f, g) is free, $\Box f, g\Box \not\equiv 0$, and for $r > 0$ the *compensation function* $m_{f,g}$ of f, g is defined

$$m_{f,g}(r) = \int_{M\langle r\rangle} \log \frac{1}{\Box f, g\Box}\sigma \ge 0. \tag{42}$$

Let $\nu : M \to \mathbb{Z}$ be a divisor with support S. Fot $t > 0$ the *counting function* n_ν of ν is defined by

(43) $$n_\nu(t) = \frac{1}{t^{2m-2}} \int_{S[t]} \nu\upsilon^{m-1} = \int_{S[t]} \nu\omega^{m-1} + n_\nu(0),$$

where the limit $n_\nu(0) = \lim_{t\to 0} n_\nu(t)$ exists. Actually since M is a vector space, $n_\nu(0) = \nu(0)$ (see Stoll [62]). For each $\mathfrak{b} \in M\langle 1\rangle$ with $j_\mathfrak{b}(\mathbb{C}) \not\subseteq S$, the pullback divisor $\nu_\mathfrak{b} = j_\mathfrak{b}^*(\nu)$ exists. If $t > 0$ then

(44) $$n_\nu(t) = \int_{\mathfrak{b}\in M\langle 1\rangle} n_{\nu_\mathfrak{b}}(t)\sigma(\mathfrak{b}).$$

Thus for $0 < s < r$ the *valence function* n_ν of ν is given by

(45) $$N_\nu(r,s) = \int_{\mathfrak{b}\in M\langle 1\rangle} N_{\nu_\mathfrak{b}}(r,s)\sigma(\mathfrak{b}) = \int_s^r n_\nu(t)\frac{dt}{t}.$$

Take reduced resprentations $\mathfrak{v} : M \to \mathbb{P}(V)$ of f and $\mathfrak{w} : M \to \mathbb{P}(V^*)$ of g. Since (f,g) is free, $h = \langle\mathfrak{v},\mathfrak{w}\rangle \not\equiv 0$. Then $\mu_{f,g} = \mu_h$ depends on f and g only. Put $S = h^{-1}(0)$. If $\mathfrak{b} \in M\langle 1\rangle$ with $j_\mathfrak{b}(\mathbb{C}) \not\subseteq S$, then $\mu_{f_\mathfrak{b},g_\mathfrak{b}} = j_\mathfrak{b}^*(\mu_{f,g})$. Hence

(46) $$N_{f,g}(r,s) = \int_{\mathfrak{b}\in M\langle 1\rangle} N_{f_\mathfrak{b},g_\mathfrak{b}}(r,s)\sigma(\mathfrak{b}) = N_{\mu_{f,g}}(r,s).$$

Thus each term in (35) is explicitely expressed.

Actually, Kneser [24] provided a more general version of (42). For $t > 0$ the *counting function* of a pure p-dimensional analytic set S in M is defined by

(47) $$n_S(t) = \frac{1}{t^{2p}} \int_{S[t]} \upsilon^p = \int_{S[t]} \omega^p + n_S(0),$$

where

(48) $$n_S(0) = \lim_{t\to 0} n_S(t)$$

exists and is called the *Lelong Number* of S at 0. Kneser assumed that $0 \notin S$, then $n_S(0) = 0$. Pierre Lelong permitted $0 \in S$ and proved (46) in 1957 [26] by the use of currents. Paul Thie [87] (1967) moved that the Lelong number is an integer. This result constituted Paul Thie's theses at Notre Dame and by coincidence Pierre Lelong was present at the defense of the theses. Of course, if $0 \in S$ then $n_S(0) > 0$. Paul Thie's result proved to be most helpful in estimating volumes from below. Of course the Lelong number of S can be defined for every $x \in M$ and shall be denoted by $L_S(x)$. Yum-Tong Siu [56] (1974) proved that the sets $\{x \in M | L_S(x) \geq q\}$ is analytic for every $q \in \mathbb{N}$. The proof was simplified by Lelong [28]

Since n_S increases, the limit

$$n_S(\infty) = \lim_{t \to \infty} n_S(t) \leq \infty \tag{49}$$

exists. As an application of value distribution theory on complex spaces, I was able to show that S is affine algebraic if and only if $n_S(\infty) < \infty$ ([63]).

This result was localized by Errett Bishop [5] (1964) to extend analytic sets over higher dimensional analytic sets. His result was refined by Shiffman [47], [48], [49].

Hellmuth Kneser did not proceed to a Second Main Theorem and a Defect Relation. Also he did not consider the possible extension of his theory to parabolic manifolds or Kähler manifolds. However, he investigated another problem: the theory of functions of finite order. He solved the two dimensional case and provided the basic ideas in m-dimensions. Later he assigned the completion of these investigations to me as my thesis topic [62], [63].

Again let (M, τ) a strictly parabolic manifold of dimensions $m > 1$. Thus M is a hermitian vector space of dimension $m > 1$ and τ is the square of the norm. If $\mathfrak{x} \in M, \mathfrak{y} \in M$, then $(\mathfrak{x}|\mathfrak{y})$ is the *hermitian product* of $\mathfrak{x}$ and $\mathfrak{y}$. If $u : \mathbb{R}_+ \to \mathbb{R}_+$ is an increasing function, its *order* is defined by

$$0 \leq \operatorname{Ord} u = \limsup_{r \to \infty} \frac{\log u(r)}{\log r} \leq \infty. \tag{50}$$

If $\nu \geq 0$ is a non-negative divisor, define $\operatorname{Ord} \nu = \operatorname{Ord} n_\nu$. Then

$\operatorname{Ord}\nu = \operatorname{Ord} N_\nu(\cdot, s)$. If $f : M \to \mathbb{P}(V)$ is a meromorphic map, define $\operatorname{Ord} f = \operatorname{Ord} T_f(\cdot, s)$.

If q is a non-negative integer, the *Weierstrass prime factor* is defined for all $z \in \mathbb{C}$ by

$$(51) \qquad E(z, q) = (1 - z)\exp(\sum_{p=1}^{q} \frac{1}{p} z^p).$$

For all $z \in \mathbb{C}(1)$ the *Kneser Kernel* is defined by

$$(52) \qquad e_m(z, q) = \frac{1}{(m-1)!} \frac{d^{m-1}}{dz^{m-1}}(z^{m-1} \log E(z, q)),$$

where $\log E(0, q) = 0$.

Let $f : M \to \mathbb{C}$ be an entire function of finite order with $f(0) = 1$. Let S be the support of the zero divisor $\nu = \mu_f$ of f. Trivially $S = f^{-1}(0)$. Assume that $S \neq \emptyset$. Then there exists a largest real number $s > 0$ such that $S(s) = 0$. Since f has finite order, there is a smallest, non-negative integer q such that

$$(53) \qquad \int_s^\infty \frac{T_f(r, s)}{r^{q+2}} dr < \infty.$$

Then $q \leq \operatorname{Ord} f \leq q + 1$. Also there exists a holomorphic function F on $M(s)$ such that $F(0) = 0$ and $f|W(s) = e^F$. By the First Main Theorem the following integral converges uniformly on every compact subset of $M(s)$ and defines a holomorphic function H on $M(s)$ with $\mu_H(0) \geq q + 1$ by

$$(54) \qquad H(\mathfrak{z}) = \int_{\mathfrak{y} \in S} \nu(\mathfrak{y}) e_m(\frac{(\mathfrak{z}|\mathfrak{y})}{(\mathfrak{y}|\mathfrak{y})}, q) \omega^{m-1}(\mathfrak{y})$$

for $\mathfrak{z} \in M(s)$. Kneser [24] shows that there is a unique polynomial P of at most degree q with $P(0) = 0$ such that

$$(55) \qquad F = P|W(s) + H \qquad f|W(s) = e^{P+H}.$$

Hence $h = e^{-P} f$ is an entire function with $\mu_h = \nu = \mu_f$ and

$h|W(s) = e^H$. Thus h depends on ν only.

Given a divisor $\nu \geq 0$ on M of finite order with $S = \operatorname{supp}\nu \neq 0$, there is a largest real number $s > 0$ such that $S(s) = 0$ and a smallest, non-negative integer q such that

$$\int_s^\infty n_\nu(t) \frac{dt}{t^{q+2}} < \infty. \tag{56}$$

Then $q \leq \operatorname{Ord}\nu \leq q + 1$. The integral (53) converges uniformly on every compact subset of $W(s)$ and defines a holomorphic function H on $M(s)$ with $H(0) = 0$ and $\mu_H(0) \geq q+1$ by (53). Does there exist an entire function h on M such that $h|W(s) = e^H$, such that $\mu_h = \nu$ and such that $\operatorname{Ord} h = \operatorname{Ord}\nu$? In his earlier paper, Kneser [23] (1936) proved the existance of such a canonical function if $m = 2$. It was my thesis problem to solve the case $m > 2$. His method required to show that a certain closed form was exact. If $m = 2$, this lead to a solvable ordinary differential equation. If $m > 2$, it took me two weeks to write out the system of partial differential equations to be solved, which I could not do. I asked him for advice. He said he had gone through the same terrible calculation and had been unable to solve the system. Then he threw away his notes. I followed his advice, but I found another proof ([62], [63]). Independently, Pierre Lelong ([25] 1953, [27] 1964) proved the existence of the canonical function h by another integral representation. Both solutions coincide by a uniquenen theorem of Rankin [42] (1968), who provided a third integral representation. In [64] 1953 I showed that the canonical function h of a 2m-periodic divisor is a theta function for this divisor and that any 2m-periodic meromorphic function is a quotient of two theta functions (Appell [2] 1891 if $m = 2$ and Poincaré [40] 1898 if $m \geq 2$). In 1975, Henri Skoda [58] and Gennadi Henkin [20] showed independently, that a non-negative divisor ν on a strictly pseudoconvex domain D in M with bounded valence N_ν is the zero divisor $\nu = \mu_h$ of a holomorphic function h on D with bounded characteristic. Later Henkin [21] (1978) showed, if $\operatorname{Ord}\nu < \infty$ then there is a holomorphic function h on D with $\nu = \mu_h$ and $\operatorname{Ord}\nu = \operatorname{Ord} h$. Recently, Polyakov [41] (1987) extended this result to the polydisc. Skoda [60] (1972) solved the problem for analytic sets of higher codimension in a complex vector space. For more details see [73].

The integral means method of Kneser fails on complex manifolds. Also he did not attempt to prove a Second Main Theorem and a Defect Relation. From the theory of holomorphic curves there are available the method of Cartan [8] and the method of Ahlfors [1] which was extended to Riemann surfaces by Hermann and Joachim Weyl [90], improved later by H. Wu [92].

In 1953/54 I extended the theory of Ahlfors-Weyl to meromorphic maps $f : M \to \mathbb{P}(V)$, where M is a m-dimensional, connected, complex manifold of dimension $m > 1$ endowed with a positive form χ of bidegree $(m-1, m+1)$ such that $d\chi = 0$. Here V is a hermitian vector space of dimension $n+1$. Again the targets are the hyperplanes in $\mathbb{P}(V)$ and f is linearly non-degenerated. Let $\mathfrak{A} = \{a_j\}_{j \in Q}$ be a family of hyperplanes $a_j \in \mathbb{P}(V^*)$ in general position. Then, under suitable assumptions a defect relation

$$\sum_{j \in Q} \delta(f, a_j) \leq n + 1 \tag{57}$$

was obtained. Also a defect relation for associated maps was proved [65]. I cannot go into details here. The extension to $m > 1$ is based on two ideas:

(1) Let $\mathfrak{G}$ be a set of open, relative compact subsets G of M with C^∞-boundary such that $\bar{g} \in G$ for all $G \in \mathfrak{G}$, where g is open with a C^∞-boundary. Assume that for each compact subset K of M there is $G \in \mathfrak{G}$ with $G \supset K$. There the Dirichlet problem $dd^c\Psi \wedge \chi = 0$ is solved for $\bar{G} - g$ with $\Psi | \partial G = 0$ and $\Psi | \partial \bar{g} = 1$.

(2) The associated maps are defined by the use of a holomorphic differential form B of bidegree $(m-1, 0)$ such that

$$0 \leq m i_{m-1} B \wedge \bar{B} \leq Y(G) \chi \qquad \text{on } \bar{G} \tag{58}$$

where $Y(G)$ is the smallest possible constant.

On parabolic manifolds the proof has been greatly simplified by Cowen-Griffiths [17] (1976), Pit-Mann Wong [93] (1976), Stoll [80] (1983), [82] (1985), [86] (1992). The definitions and identities (34)–(46) also hold on parabolic manifolds except, of course, for the slicing $j_{\mathfrak{b}}$ and the equality $n_\nu(0) = \nu(0)$ and (41) may be vacuous, since f may not have a global, reduced representation on M. The defect of (f, g) is defined as in (16). For an exact statement of the defect relation I refer to the papers mentioned before, but I will state the defect relation in a special case with a new variation:

Let M be a connected, complex manifold of dimension $m > 1$. Let W be a hermitian vector space of dimension m. Let $\pi : M \to W$ be a surjective, proper, holomorphic map. Then $\tau = \|\pi\|^2$ is a parabolic exhaustion of M and (M, τ) is called *a parabolic covering space of* W. Take any holomorphic form ζ of bidegree $(m, 0)$ on W without zeroes. Then the zero divisor $\beta \geq 0$ of $\pi^*(\zeta)$ does not depend on the choice of ζ and is called the *branching divisor* of π. Put $B = \text{supp}\beta$. Then π is locally biholomorphic at $z \in M$ if and only if $z \in M - B$. Since π is proper and holomorphic, $B' = \pi(B)$ and $\hat{B} = \pi^{-1}(B')$ are analytic and $\pi : M - \hat{B} = W - B'$ is a covering space in the sense of topology. Its sheet number ς is given by (31).

Let V be a hermitian vector space of dimension $n + 1 > 1$. Let $f : M \to \mathbb{P}(V)$ be a linearly non-degenerated meromorphic map of transcendental growth (i.e. $A_f(\infty) = \infty$). Assume that the *Ricci defect*

$$R_f = \lim_{r\to\infty} \frac{N_\beta(r, s)}{T_p(r, s)} < \infty. \tag{59}$$

Let $\mathfrak{A} = \{a_j\}_{j\in Q}$ a finite family of hyperplanes $a_j \in \mathbb{P}(V^*)$ in general position. Then we have the **Defect Relation**

$$\sum_{j\in Q} \delta(f, a_j) \leq n + 1 + \frac{1}{2} n(n+1) R_f. \tag{60}$$

A meromorphic map $h : M \to \mathbb{P}(V)$ is said to *separate the fibers of* π, if there is a point $x \in W - B'$ such that $\pi^{-1}(x) \wedge I_h = \emptyset$ and such that $h|\pi^{-1}(x)$ is injective. If so, and if $s > 0$, there is a constant $C(s) > 0$ such that

$$N_\beta(r, s) \leq 2(\varsigma - 1) T_h(r, s) + C(s) \tag{61}$$

for all $r > 0$ (Noguchi [38], Stoll [83]). Define

(62) $\mathfrak{h} = \bigcup_{k \in \mathbb{N}} \{h | h : M \to \mathbb{P}_k \text{ meromorphic, separates fibers of } \pi\}$

Then the separation index of f is defined by

$$\gamma = \inf_{h \in \mathfrak{h}} \limsup_{r \to \infty} \frac{T_h(r, s)}{T_f(r, s)}. \tag{63}$$

If f separates the fibers of π, then $\gamma \leq 1$. We obtain the **Defect Relation**

$$\sum_{j \in Q} \delta(f, a_j) \leq n + 1 + n(n+1)(\varsigma - 1)\gamma \tag{64}$$

If $n = 1$, that is, if f is a meromorphic function with transcendental growth separating the fibers of π, then

$$\sum_{j \in Q} \delta(f, a_j) \leq 2\varsigma \tag{65}$$

which, in the case $m = 1$, was already proved by H. Cartan [8] (1933).

In 1977 Al Vitter [89] proved the Lemma of the logarithmic derivative for meromorphic functions on a hermitian vector space W and derived the defect relation for meromorphic maps $f : W \to \mathbb{P}(V)$ by Cartan's original method. For a detailed account see also Stoll [79], 1982. E. Bardis [3] (1990) extended the result to parabolic covering spaces of W.

In 1973–74, Carlson and Griffiths [16] and Griffiths and King [19] invented a new method to prove the defect relation. In keeping within [19], the result shall be stated only in the case of a parabolic covering space (M, τ) of a hermitian vector space of dimension $m > 1$. The advantage of the new method is, that it applies to holomorphic maps $f : M \to N$, where N is a connected, n-dimensional, compact, complex manifold. A positive holomorphic line bundle L spanned by its holomorphic sections is given on N. Then N is projective algebraic. The disadvantage of the new method is, that we have to assume that the map f is *dominant* which means that rank $f = n$. The vector space Y^* of all holomorphic sections of L have finite dimension $k + 1 > 1$. If $0 \neq \mathfrak{a} \in Y^*$, the zero set $E_L[a] = \{x \in N | \mathfrak{a}(x) = 0\}$ depends on

$a = \mathbb{P}(\mathfrak{a}) \in \mathbb{P}(Y^*)$ only. Let $Y = (Y^*)^*$ be the dual vector space of Y. If $x \in N$, the linear subspace $\Phi(x) = \{\mathfrak{a} \in V^* | \mathfrak{a}(x) = 0\}$ has dimension k. Thus one and only one $\varphi(x) \in \mathbb{P}(Y)$ exists such that $E[\varphi(x)] = \mathbb{P}(\Phi(x))$. The holomorphic map $\varphi : N \to \mathbb{P}(Y)$ is called the *dual classification map of* L. The value distribution functions of f are defined as those of $\varphi \circ f$. First Main Theorem holds but the defect relation so obtained is not optimal. As before we assume that f has transcendental growth and that there is given a finite family $\mathfrak{A} = \{a_j\}_{j \in Q}$ of points $a_j \in \mathbb{P}(Y^*)$. However we have to consider the geometry of $\{E_L[a_j]\}_{j \in Q}$ and not the geometry of $\{E[a_j]\}_{j \in Q}$. Define

(66) $$E_L[\mathfrak{A}] = \bigcup_{j \in Q} E_L[a_j].$$

For each $j \in Q$ take $\mathfrak{a}_j \in V_*^*$ with $a_j = \mathbb{P}(\mathfrak{a}_j)$. Take $x \in E_L[\mathfrak{A}]$. Then

(67) $$P = \{j \in Q | x \in E_L[a_j]\} = \{j \in Q | \mathfrak{a}_j(x) = 0\} \neq \emptyset$$

Put $p = \# P$. Take a bijective map $\lambda : \mathbb{N}[1, p] \to P$. There is an open, connected neighborhood U of x and a holomorphic section $\mathfrak{b} : U \to L$ such that $\mathfrak{b}(z) \neq 0$ for all $z \in U$. For each $j \in \mathbb{N}[1, p]$, there is one and only one holomorphic function h_j on U such that $\mathfrak{a}_{\lambda(j)} | U = h_j \mathfrak{b}$. Then $\mathfrak{A}$ is said to have *strictly normal crossings at* x if and only if

(68) $$dh_1(x) \wedge \ldots \wedge dh_p(x) \neq 0.$$

The definition is independent of the choices which were made. $\mathfrak{A}$ is said to have *strictly normal crossings* if $\mathfrak{A}$ has strictly normal crossings at every $x \in E_L[\mathfrak{A}]$, which we assume now.

Let K be the canonical bundle of N Let K^* be the dual bundle to K. Define

(69) $$\left[\frac{K^*}{L}\right] = \inf\{\frac{v}{w} | v \in \mathbb{N}, w \in \mathbb{N}, L^v \otimes K^w \text{ positive } \}.$$

Define R_f by (59) and γ by (63). With these assumptions and definitions, the

Defect Relation of Griffiths-King

$$\sum_{j\in Q} \delta(f, a_j) \leq \left[\frac{K^*}{L}\right] + R_f \tag{70}$$

$$\sum_{j\in Q} \delta(f, a_j) \leq \left[\frac{K^*}{L}\right] + 2(\varsigma - 1)\gamma \tag{71}$$

holds. In [75] (1977) the theory was refined and extended to general parabolic manifolds.

A difficult, major, unsolved problem is the question if "dominant" can be replaced by another assumption which does not imply $m \geq n$. For instance does (70) hold if $f(M)$ is not contained in any proper analytic subset of N? As Biancofiore [4] has shown the assumption $f(M) \not\subseteq E_L[a]$ for all $a \in \mathbb{P}(Y^*)$ does not suffice. Can the condition "strictly normal crossings" be relaxed?

Let V be a hermitian vector space of dimension $n+1 > 1$. Apply the previous theory to $N = \mathbb{P}(V)$. Let H be the hyperplane section bundle on $\mathbb{P}(V)$. Take $p \in \mathbb{N}$ and choose $L = H^p$. Then $K = H^{-n-1}$ and $L^v \otimes K^w = H^{pv-w(n+1)}$. Thus $[\frac{K^*}{L}] = \frac{n+1}{p}$. Thus (70) and (71) reads

$$\sum_{j\in Q} \delta(f, a_j) \leq \frac{n+1}{p} + R_f \tag{72}$$

$$\sum_{j\in Q} \delta(f, a_j) \leq \frac{n+1}{p} + 2(\varsigma - 1)\gamma. \tag{73}$$

If $p = 1$, this is sharper than (60) which is due to the dominance of f.

Until now, target families of codimension 1 only where considered. Does there exist a value distribution theory for codimension $\ell > 1$. In 1958, H. Levine [30] proved an unintegrated First Main Theorem for projective planes of codimension $\ell > 1$ in $\mathbb{P}(V)$. At the 1958 Summer School at the University of Chicago, S. S. Chern asked me to find the integrated version. When I left, I told him that there is no such thing. I was much surprised when he published an integrated version [10] (1960) shortly afterwards. I failed, since I insisted on an old version to be obtained and because I had forgotten one of Max Planck's admonitions in one of his textbooks: "The energy principle is not a law of nature, but of man. Each time it fails in nature, man invents a new type of energy to restore the principle." The First Main Theorem is such a principle. In order to retain it, S. S. Chern had to admit a new, nasty term, later called the *deficit*, into the equation.

In 1965, Bott and Chern [6] extended the First Main Theorem to the equidistribution of the zeroes of holomorphic sections in hermitian vector bundles. Thus differential geometry was brought into value distribution theory. Later the theory was expanded to include all Schubert varieties associated to holomorphic vector bundles. With the work of H. Wu [91] (1968–70), F. Hirschfelder [22] (1969), L. Dektjarev [18] (1970), Michael Cowen [16] (1973), Chia-Chi Tung [88] (1973), and myself [67] (1967) [68] (1969) [69] (1970) and [76] (1978) a wide range of First Main Theorems for codimension $\ell > 1$ was established.

Mostly, they can be brought under the following scheme

(74)

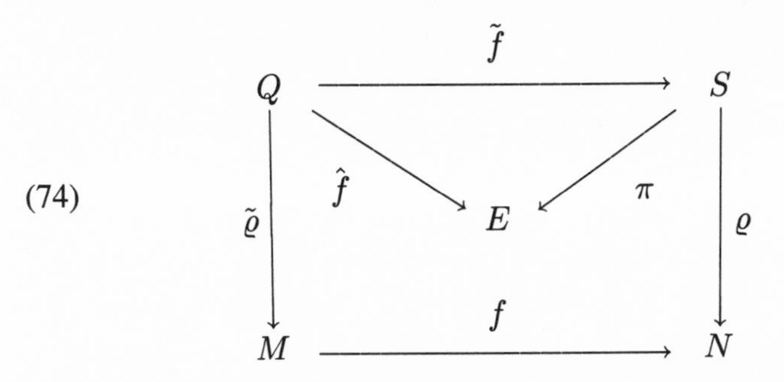

Where M, N and E are connected, complex manifolds of dimensions m, n and k respectively. Here E is a compact Kähler manifold and S is an analytic subset of $N \times E$. The projections ϱ and π are surjective, open and of pure fiber dimensions q and p respectively with $n - p = \ell \geq 1$ and $m - \ell \geq 0$. The map ϱ is locally a product at every point of S. Since E is compact, ϱ is proper. Thus

(75) $$\dim S = p + k = n + q \qquad k - q = n - p = \ell.$$

The diagram is completed as a pull back by the holomorphic map f:

(76) $$Q = \{(x, z) \mid f(x) = \varrho(z)\}$$

(77) $\tilde{\varrho}(x,z) = x \qquad \tilde{f}(x,z) = 1 \qquad \hat{f}(x,z) = \pi(z)$

(78) $\varrho \circ \tilde{f} = f \circ \tilde{\varrho} \qquad \hat{f} = \tilde{f} \circ \pi.$

The map $\tilde{\varrho}$ has pure fiber dimensions q and is locally a product at every point of Q. Hence Q has pure dimension $m+q$.

For each $y \in E$, the analytic subset $S_y = \varrho(\pi^{-1}(y))$ is a pure p-dimensional analytic subset of N. The family $\mathfrak{S} = \{S_y\}_{y \in E}$ is the target family for the holomorphic map f. We assume that $E_y = f^{-1}(S_y)$ is either empty or has generically the dimension $m - \ell$. Let $\xi > 0$ be the Kähler volume for of E with

(79) $$\int_E \xi = 1$$

Let ϱ_* be the fiber integration operator. Then $\Omega = \varrho_* \pi^*(\xi)$ is a non-negative closed form of bidegree (ℓ, ℓ) and class C^∞ on N. Here Ω is the Poincaré dual of the homology class defined by $\mathfrak{S}$. Take $y \in E$, by Hodge theory or construction (H. Wu [91], Stoll [69]) there is a non-negative form $\lambda_y \geq 0$ on $E - \{y\}$ of bidegree $(k-1, k-1)$ with residue 1 at y such that

(80) $$dd^c \lambda_y = \xi \qquad \text{on } E - \{y\}$$

Then $\Lambda_y = \varrho_* \pi^*(\lambda_y) \geq 0$ is a form of bidegree $(\ell - 1, \ell - 1)$ on $N - S_y$ with

(81) $$dd^c \Lambda_y = \Omega \qquad \text{on } N - S_y$$

Let φ be a form of bidegree $(m-\ell, m-\ell)$ and of class C^∞ with compact support in M. With proper multiplicities ν_y, the Stokes Theorem, the Residue Theorem and fiber integration imply

(82) $$\int_M f^*(\Lambda_y) \wedge dd^c \varphi = -\int_M df^*(\Lambda_y) \wedge d^c \varphi$$
$$= -\int_M d\varphi \wedge d^c f^*(\Lambda_y)$$

$$= -\int_{F_y} \nu_y \varphi + \int_M \varphi \wedge dd^c f^*(\Lambda_y),$$

if E_y has pure dimension $m - \ell$. As a generalization of the Poincaré-Lelong formula we obtain the **Unintegrated First Main Theorem**

$$\int_M f^*(\Omega) \wedge \varphi = \int_{F_y} \nu_y \varphi + \int_M f^*(\Lambda_y) \wedge dd^c \varphi. \tag{83}$$

For the integration, we assume that an exhaustion $\tau : M \to \mathbb{R}_+$ is given with

$$w = dd^c \log \tau \geq 0 \quad \upsilon = dd^c \tau \geq 0 \quad \sigma_\ell = d^c \log \tau \wedge w^{m-\ell}, \tag{84}$$

Then $d\sigma_\ell = w^{m-\ell+1}$. We keep the notations (27) and (28), but do not require that τ is parabolic. For $t > 0$ the *spherical image function* is defined by

$$A_f(t) = \frac{1}{t^{2m-2\ell}} \int_{M[t]} f^*(\Omega) \wedge \upsilon^{m-\ell} \geq 0. \tag{85}$$

For $0 < s < r$ the *characteristic function* is defined by

$$T_f(r, s) = \int_s^r A_f(t) \frac{dt}{t} \geq 0. \tag{89}$$

Take $y \in E$ such that E_y has pure codimension ℓ or is empty. For all $t > 0$ the *counting function* is defined by

$$n_{f,y}(t) = \frac{1}{t^{2m-2\ell}} \int_{E_y[t]} \nu_y \upsilon^{m-\ell} \geq 0 \tag{90}$$

and for $0 < s < r$ the *valence function* is defined by

$$N_{f,y}(r, s) = \int_s^r n_{f,y}(t) \frac{dt}{t} \geq 0. \tag{91}$$

For almost all $r > 0$ the *compensation function* is defined by

$$m_{f,y}(r) = \frac{1}{2} \int_{M\langle r\rangle} f^*(\Lambda_y) \wedge \sigma_\ell \geq 0. \tag{92}$$

For $0 < s < r$ the *deficit* is defined by

$$D_{f,y}(r,s) = \frac{1}{2} \int_{M[r]-M[s]} f^*(\Lambda_y) \wedge \omega^{m-\ell+1}. \tag{93}$$

If $\ell = 1$ and τ is parabolic, then $\omega^m \equiv 0$ which implies $D_{f,y} \equiv 0$. However if $\ell > 1$, then this is false even if τ is parabolic. The same calculation as in (82) but respecting boundary terms yields the **First Main Theorem**

$$T_f(r,s) = N_{f,y}(r,s) + m_{f,y}(r) - m_{f,y}(s) - D_{f,y}(r,s). \tag{94}$$

A continuous form $\hat{\lambda} \geq 0$ bidegree $(k-1, k-1)$ on E exists such that $x \in E$ implies

$$\hat{\lambda}(x) = \int_{y \in E} \lambda_y(x) \otimes \xi(y) \geq 0. \tag{95}$$

The $\hat{\Lambda} = \varphi_* \pi^*(\hat{\lambda}) \geq 0$ is a continuous form of bidegree $(\ell-1, \ell-1)$ on N. For all $x \in N$, fiber integration yields

$$\hat{\Lambda}(z) = \int_{y \in E} \Lambda_y(z) \otimes \xi(y) \geq 0. \tag{96}$$

Thus we obtain

$$\mu_f(r) = \int_{y \in E} m_{f,y}(r)\xi(y) = \frac{1}{2} \int_{M\langle r\rangle} f^*(\hat{\Lambda}) \wedge \sigma_\ell \geq 0 \tag{97}$$

$$\Delta_f(r,s) = \int_{y \in E} D_{f,y}(r,s)\xi(y) = \frac{1}{2} \int_{M[r]-M[s]} f^*(\hat{\Lambda}) \wedge \omega^{m-\ell-1} \geq 0 \tag{98}$$

$$T_f(r,s) = \int_{y\in E} N_{f,y}(r,s)\xi(y) \geq 0, \tag{99}$$

which implies

$$\Delta_f(r,s) = \mu_f(r) - \mu_f(s). \tag{100}$$

For $r > 0$ define

$$B(r) = \{y \in E \mid E_y \cap M[r] \neq \emptyset\} \tag{101}$$

$$0 \leq b_f(r) = \int_{B(r)} \xi \leq 1$$

$$B = \{y \in E \mid E_y \neq \emptyset\} \tag{102}$$

$$0 \leq b_f = \int_B \xi \leq 1.$$

Then $B = \bigcup_{r>0} B(r)$ and $b_f(r) \to b_f$ for $r \to \infty$ increasingly. Now (94) implies

$$N_{f,y}(r,s) \leq T_f(r,s) + m_{f,y}(s) + D_{f,y}(r,s). \tag{103}$$

If $y \in E - B(r)$, then $N_{f,y}(r,s) = 0$ and (99) implies

$$\begin{aligned} T_f(r,s) &= \int_{y\in E} N_{f,y}(r,s)\xi(y) = \int_{y\in B(r)} N_{f,y}(r,s)\xi(y) \\ &\leq \int_{y\in B(r)} (T_f(r,s) + m_{f,y}(s) + D_{f,y}(r,s))\xi(y) \\ &\leq b_f(r)T_f(r,s) + \int_{y\in E} (m_{f,y}(s) + D_{f,y}(r,s))\xi(y) \\ &= b_f(r)T_f(r,s) + \mu_f(s) + \Delta_f(r,s). \end{aligned} \tag{104}$$

Therefore

$$0 \leq (1 - b_f(r)) \leq \frac{\mu_f(s) + \Delta_f(r,s)}{T_f(r,s)} \qquad \text{if } r > s > 0. \tag{105}$$

Assume that $T_f(r,s) \to \infty$ and $\Delta_f(r,s)/T_f(r,s) \to 0$ for $r \to \infty$. Then $b_f = 1$. Thus $f(M)$ intersects almost all targets S_y. Even for holomorphic curves on $\mathbb{C}$ surprising results can be obtained:

Proposition

A holomorphic map $f : \mathbb{C} \to \mathbb{P}_6$ is defined for all $z \in \mathbb{C}$ by

$$f(z) = \mathbb{P}(1, e^{\zeta z}, e^{\zeta^2 z}, \ldots, e^{\zeta^6 z}) \tag{106}$$

where $\zeta = e^{\frac{\pi i}{3}}$. If $r \geq \frac{\pi}{6}(245 + \log 7) \approx 129.3006$, then $f(\mathbb{C}[r])$ intersects at least 99% of all hyperplanes in $\mathbb{P}_6$.

Proof. A reduced representation $\mathfrak{v}$ of f is defined for all $z \in \mathbb{C}$ by

$$\mathfrak{v}(z) = (1, e^{\zeta z}, e^{\zeta^2 z}, \ldots, e^{\zeta^6 z})$$

with $\mathfrak{v}(0) = (1, \cdots, 1)$. Thus $\|\mathfrak{v}(0)\| = \sqrt{7}$. We can take $s = 0$. Thus

$$T_f(r,0) = \int_{\mathbb{C}\langle r\rangle} \log \|\mathfrak{v}\| \sigma - \frac{1}{2} \log 7$$

Observe that

$$L = \sum_{j=1}^{6} |\zeta^j - \zeta^{j-1}| = 6.$$

By Stoll [80] Proposition 15.5 page 201 we have

$$0 \leq \int_{\mathbb{C}\langle r\rangle} \log \|\mathfrak{v}\| \sigma - \frac{L}{2\pi} r \leq \frac{1}{2} \log 7.$$

Thus

$$\frac{6r - \pi \log 7}{2\pi} \leq T_f(r,0).$$

By Stoll [80] (6.66) page 140 we have $\mu_f(s) = \frac{1}{2}\sum_{p=1}^{6} \frac{1}{p} = \frac{49}{40}$ for all $s > 0$. If $r > (\pi/6)\log 7$, then

$$0 \leq 1 - b_f(r) \leq \frac{49}{40} \frac{2\pi}{6r - \pi \log 7} = \frac{49}{20} \frac{\pi}{6r - \pi \log 7}$$

Define $r_0 = \frac{\pi}{6}(245 + \log 7)$. Take $r \geq r_0$ Then

$$0 \leq 1 - b_f(r) \leq \frac{49}{20} \frac{\pi}{6r_0 - \pi \log 7} = \frac{49}{20 \times 245} = \frac{1}{100}.$$

Hence $b_f(r) \geq \frac{99}{100}$, q.e.d.

This calculation was made possible by a theorem of Shiffman-Weyl. The method can be greatly improved, see Molzon, Shiffman, and Sibony [31] (1981), and Lelong and Gruman [29] (1986).

In 1929 Rolf Nevanlinna [33] conjectured that his defect relation remains valid, if the constant target points $a_j \in \mathbb{P}_1$ are replaced by "target" functions $g_j : \mathbb{C} \to \mathbb{P}_1$ which move slower than the "hunter" function $f : \mathbb{C} \to \mathbb{P}_1$, that is, if

$$(107) \qquad T_{g_j}(r, s)/T_f(r, s) \to 0 \qquad \text{for } r \to \infty.$$

In 1964 Chi-Tai Chuang [14] proved the conjecture for entire functions $f : \mathbb{C} \to \mathbb{C}$ and created the basis for the solution of the problem. In 1986, Norbet Steinmetz [61] proved Nevanlinna's conjecture. In 1991, Ru Min and I [43] [44] [85] proved the conjecture for holomorphic curves and solved the case of the Cartan conjecture for moving targets [46]. In 1985, Charles F. Osgood [39] claimed that these theorems are a consequence of his results in diophantine approximation, but to me this implication is not self evident and still has to be established.

At the end let me state a result at Notre Dame on this subject matter, combining the work of Emmanuel Bardis [3], and Ru Min and myself [44].

At first some concepts have to be explained. Let M be a connected, complex manifold of dimension m. Let V be a hermitian vector space of finite dimension $n + 1 > 1$. Let $f : M \to \mathbb{P}(V)$ be a meromorphic map. Take $\mathfrak{a} \in V^*$ and $0 \neq \mathfrak{b} \in V^*$. Put $b = \mathbb{P}(\mathfrak{b})$. Assume that (f, b) is free. Then there exists one and only one meromorphic function $f_{\mathfrak{a},\mathfrak{b}}$ on M, called *a coordinate function*, such that for each point $p \in M$ there exists an open, connected neighborhood U of p and a reduced representation $\mathfrak{v} : U \to V$ such that

(108) $$f|U = \frac{<\mathfrak{v}, \mathfrak{a}>}{<\mathfrak{v}, \mathfrak{b}>}.$$

Here $<\mathfrak{v}, \mathfrak{b}> \not\equiv 0$ since (f, b) is free. Let $\mathfrak{C}_f$ be the set of all those coordinate functions of f. Trivially $\mathbb{C} \subseteq \mathfrak{C}_f$. Let $\mathfrak{M}$ be the field of meromorphic functions on M. Let $\mathfrak{K}$ be a subfield of $\mathfrak{M}$. The f is said to be *defined over* $\mathfrak{K}$ if and only if $\mathfrak{C}_f \subseteq \mathfrak{K}$. The meromorphic map f is said to be *linearly non-degenerated* over $\mathfrak{K}$ if and only if (f, g) is free for every meromorphic map $g : M \to \mathbb{P}(V^*)$ defined over $\mathfrak{K}$. Let $\mathfrak{G} = \{g_j\}_{j \in Q}$ be a finite family of meromorphic maps $g_j : M \to \mathbb{P}(V^*)$ with indeterminacy I_{g_j}. Define

(109) $$I_{\mathfrak{G}} = \bigcup_{j \in Q} I_{g_j} \qquad \mathfrak{C}_{\mathfrak{G}} = \bigcup_{j \in Q} \mathfrak{C}_{g_j}.$$

Let $\mathfrak{K}_{\mathfrak{G}} = \mathbb{C}(\mathfrak{C}_{\mathfrak{G}})$ be the extension field of $\mathfrak{C}$ in M generated by $\mathfrak{C}_{\mathfrak{G}}$. The family $\mathfrak{G}$ is said to be in *general position* if and only if there is a point $z \in M - I_{\mathfrak{G}}$ such that $\mathfrak{G}(z) = \{g_j(z)\}_{j \in Q}$ is in general position.

Theorem: Defect relation for moving target.

Let M be a connected, complex manifold of dimension M. Let W be a hermitian vector space of dimension m. Let $\pi : M \to W$ be a surjective, proper holomorphic map. Then $\tau = \|\pi\|^2$ is a parabolic exhaustion of M. Let V be a hermitian vector space of finite dimension $n + 1 > 1$. Let $\mathfrak{G} = \{g_j\}_{j \in Q}$ be a finite family of meromorphic maps $g : M \to \mathbb{P}(V^*)$ in general position. Assume at least on $k \in Q$ exists such that g_k is not constant and separates the fibers of π. Let $f : M \to \mathbb{P}(V)$ be a meromorphic map which is linearly non-degenerated over $\mathfrak{K}_{\mathfrak{G}}$. Assume that g_j grows slower than f for each $j \in Q$. Then

(110) $$\sum_{j \in Q} \delta(f, g_j) \leq n + 1.$$

During the time from 1933 to 1960 the foundation was laid. The 1960th was the decade of the First Main Theorem. The 1970th was the decade of the Second Main Theorem. The 1980th was the decade of the moving targets. Perhaps the 1990th will be a decade of refinement and of value distribution over function fields in conjunction with diophantine approximation.

References

[1] Ahlfors, L., *The theory of meromorphic curves*. Acta. Soc. Sci. Fenn. Nova Ser. A **3** (4) (1941) 171–183.

[2] Appell, P., *Sur les fonctions périodiques de deux variables*, J. Math. Pures Appl. (4) **7** (1891), 157–219.

[3] Bardis, E., *The Defect Relation for Meromorphic Maps Defined on Covering Parabolic Manifolds*, Notre Dame Thesis (1990), pp. 133.

[4] Biancofiore, A. *A hypersurface defect relation for a class of meromorphic maps*, Trans. Amer. Math. Soc. **270** (1982), 47–60.

[5] Bishop, E., *Condition for the analycity of certain sets*, Duke Math. J. **36** (1969), 283–296.

[6] Bott, R. and Chern, S. S. *Hermitian vector bundles and the equidistribution of the zeroes of their holomorphic sections*. Acta Math. **114** (1965), 85–121.

[7] Carlson, J. and Griffiths, Ph., *A defect relation for equidimensional holomorphic mappings between algebraic varieties*, Ann. of Math. (2) **95** (1972), 557–584.

[8] Cartan, H., *Sur les zéros des combinaisons linéaires de p fonctions holomorphes données*. Mathematica (Cluj) **7** (1933) 80–103.

[9] Chen, W., *Cartan conjecture: Defect relation for meromorphic maps from manifold to projective space*. Notre Dame Thesis (1987) pp. 166.

[10] Chern, S. S. *The integrated form of the first main theorem for complex analytic mappings in several variables*. Ann. of Math. **77** (1960), 536–551.

[11] ———. *Complex analytic mappings of Riemann surfaces*. I. Amer. J. Math. **82** (1960), 323–337.

[12] ———. *Holomorphic curves in the plane, in "Differential*

Geometry in honor of K. Yano". Kinokuniya, Tokyo, (1972), 72–94.

[13] ______. *On holomorphic mappings of hermitian manifolds of the same dimension*. Proc. Symp. Pure Math. **11** (1968). Entire Functions and Related Parts of Analysis, Amer. Math. Soc., 157–170.

[14] Chuang, Ch. T. *Une généralization d'une inégalité de Nevanlinna*. Sci. Sin. **13** (1964), 887–895.

[15] ______. *On the zeros of some differential polynomials of meromorphic functions*. Science Report **89-002**, Inst. of Math. Peking University (1989), 1–29.

[16] Cowen, M. *Hermitian vector bundles and value distribution for Schubert cycles*. Trans. Amer. Math. Soc. (180) (1973), 189–228.

[17] Cowen, M. and Griffiths, Ph., *Holomorphic curves and metrics of non-negative curvature*. J. Analyse Math. **29** (1976) 93–153.

[18] Dektjarev, L. *The general first fundamental theorem of value distribution*. Dokl. Akad. Nauk. SSR **193** (1970) (Soviet Math. Dokl. **11** (1970), 961–63).

[19] Griffiths, Ph. and King, J., *Nevanlinna theory and holomorphic mappings between algebraic varieties*, Acta Math. **130** (1973), 145–220.

[20] Henkin, G. H., *Solutions with estimates of the H. Lewy and Poincaré-Lelong equations. Constructions of functions of the Nevanlinna class with prescribed zeroes in strictly pseudoconvex domains*, Dokl. Akad. Nauk SSSR **210** (1975), 771–774 (Soviet Math. **16** (1975), 1310–1314).

[21] Henkin, G. M., and Dautov, S. A., *Zeroes of holomorphic functions of finite order and weighted estimates for the solutions of the* $\bar{\partial}$*-equation*, Mat. Sb. (N.S.) **107** (**149**) (1978), 163–174, 317.

[22] Hirschfelder, J. *The first main theorem of value distribution in several variables*. Invent. Math. **8** (1969), 1–33.

[23] Kneser, H., *Ordnung und Nullstellen bei ganzen Funktionen zweier Veränderlicher*, S.-B. Press Akad. Wiss. Phys.-Math. Kl. **31** (1936), 446–462.

[24] ______, *Zur Theorie der gebrochenen Funktionen mehrerer Veränderlicher*, Jber, Deutsch. Math. Verein **48** (1938), 1–38.

[25] Lelong, P., *Sur l'extension aux fonctions entières de n variables, d'ordre fini, a'un development canonique de Weierstrass*, CR Acad. Sei., Paris, **237** (1953), 865–867.

[26] ______, *Intégration sur une ensemble analytique complexe*, Bull. Soc. Math. France **85** (1957), 328–370.

[27] ______, *Fonctions entières (n-variables) et fonctions plurisousharmoniques d'order fini dans $\mathbb{C}^n$*, J. Analyse Math. **12** (1964) 365–407.

[28] Lelong, P., *Sur la structure des courants positif's fermés*, Lecture Notes in Mathematics **578** (1977) 136–158.

[29] Lelong, P. and Gruman, L. *Entire Functions of Several Complex Variables*. Grundl d. Wiss. **282** (1986) pp. 270, Springer-Verlag.

[30] Levine, H. *A theorem on holomorphic mappings into complex projective space*. Ann. of Math. **71** (1960), 529–535.

[31] Molzon, R. E., Shiffman, B., and Sibony, N. *Average growth estimates for hyperplane sections of entire analytic sets*. Math. Ann. **257** (1981), 43–53.

[32] Nevanlinna, R., *Zur Theorie der meromorphen Funktionen*, Acta Mathematica **46** (1925) 1–99.

[33] ______. *Le Théorème de Picard-Borel et la Théorie des Fonctions Meromorphes*, Gauthiers-Villars, Paris (1929) reprint Chelsea Publ. Co. New York (1974) pp. 171.

[34] ______. *Eindeutige analytische Funktionen* 2nd *ed.* Die Grundl d. Math Wiss. **46** (1953) pp. 379. Springer-Verlag.

[35] Nochka, E. I., *Defect relations for meromorphic curves*. Izv, Akad. Nauk. Moldav. SSR Ser. Fiz. Teklam. Mat. Nauk. (1982), 41–47.

[36] ______, *On a theorem from linear algebra Izv.*. Akad. Nauk. Modav. SSR Ser. Fiz. Teklam Mat. Nauk. (1982) 29–33.

[37] ______, *On the theory of meromorphic curves*. Dokl, Akad. Nauk. SSR (1983), 377–381.

[38] Noguchi, J., *Meromorphic mappings of a covering space*

over $\mathbb{C}^n$ *into projective algebraic variety and defect relations*, Hiroshima Math. J. **6** (1976), 265–280.

[39] Osgood, Ch. F. *Sometimes effective Thue-Siegel-Roth-Schmidt-Nevanlinna bounds or better*. J. Number Theory **21** (1985), 347–389.

[40] Poincaré, H., *Sur les propriétiés du potential algébriques*, Acta. Math. **22** (1898), 89–178.

[41] Polyakov, P., *Zeroes of holomorphic functions of finite order in a polydisk*, Mat. Sb. (N.S.) **133** (175) (1987), 103–111, 114.

[42] Ronkin, L. I., *An analog of the canonical product for entire functions of several complex variables*, Trudy Moskov Mat. Obšč. **18** (1968), 105–146 = Trans. Moscow Math. Soc. **18** (1968), 117–160.

[43] Ru, M. and Stoll, W. *Courbe holomorphes évitant des hyperplans mobiles*. C. R. Acad. Sci. Paris **310** Série I (1990), 45–48.

[44] ———. *The Second Main Theorem for Moving Targets*. J. Geom. Anal. **1** (1991), 99–138.

[45] ———. *The Nevanlinna Conjecture for moving targets*, preprint pp. 16.

[46] ———. *The Cartan Conjecture for Moving Targets*. Proceedings of Symposia in Pure Mathematics. **52** (1991) 477–508.

[47] Shiffman, B., *On the removal of singularities of analytic sets*, Michigan Math. J. **15** (1968), 111–120.

[48] ———, *On the continuation of analytic curves*, Math. Ann. **184** (1970), 268–274.

[49] ———, *On the continuation of analytic sets*, Math. Ann. **185** (1970), 1–12.

[50] ———, *Nevanlinna defect relations for singular divisors*, Invent. Math. **31** (1975), 155–182.

[51] ———, *Holomorphic curves in algebraic manifolds*, Bull. Amer. Math. Soc. **83** (1977), 553–568.

[52] ———, *On holomorphic curves and meromorphic maps in projective spaces*, Indiana Univ. Math. J. **28** (1979), 627–641.

[53] ______, *Introduction to Carlson-Griffiths equidistribution theory*, Lecture Notes in Mathematics, **981** (1983), 64–89, Springer-Verlag.

[54] ______, *New defect relations for meromorphic functions on* $\mathbb{C}^n$, Bull. Amer. Math. Soc. (New Series) **7** (1982), 594–601.

[55] ______, *A general second main theorem for meromorphic functions on* $\mathbb{C}^n$, Amer. J. Math. **106** (1984), 509–531.

[56] Siu, Y. T., *Analyticity of sets associated to Lelong numbers and the extension of closed positive currents*, Invent. Math. **27** (1974), 53–156.

[57] Skoda, H., *Croissance des fonctions entiéres s'annulant sur une hypersurface donnée de* $\mathbb{C}^n$, Seminair P. Lelong 1970–71, Lecture Notes in Mathematics **275** (1972), 82–105, Springer-Verlag.

[58] ______, *Valeurs au board les solutions de l'operateur d", et caracterisation des zéros des fonctions de la classe Nevanlinna*, Bull. Soc. Math. France **104** (1976), 225–299.

[59] ______, *Solution à croissance du second problème Cousin daus* $\mathbb{C}^n$, Ann. Inst. Fourier (Grenoble) **21** (1971), 11–23.

[60] ______, *Sous-ensembles analytiques d'ordre fini ou infini daus* $\mathbb{C}^n$, Bull. Soc. Math. France **100** (1972), 353–408.

[61] Steinmetz, N. *Eine Verallgemeinerung des zweiten Nevanlinnaschen Hauptsatzes*. J. Reine Angew. Math. **368** (1986), 134–141.

[62] Stoll, W. *Mehrfache Integrale auf komplexen Mannigfaltigkeiten*. Math. Zeitschr. **57** (1952), 116–154.

[63] ______. *Ganze Funktionen endlicher Ordnung mitgegebenen Nullstellenflächen*. Math. Zeitschr. **57** (1953), 211–237.

[64] ______. *Konstruktion Jacobischer und mehrfach periodischer Funktionen zu gegebenen Nullstellenflächen*. Math. Zeitschr. **126** (1953), 31–43.

[65] ______. *Die beiden Hauptsätze der Wertverteilungstheorie bei Funktionen mehrerer komplexen Veränderlichen*. I Acta Math. **90** (1953), 1–115, II Acta Math. **92** (1954), 55–169.

[66] ______. *The growth of the area of a transcendental analytic set*. I, II Math. Ann. **156** (1964), 47–78, 144–170.

[67] ———. *A general first main theorem of value distribution.* Acta Math. **118** (1967), 111–191.

[68] ———. *About the value distribution of holomorphic maps into projective space.* Acta Math. **123** (1969), 83–114.

[69] ———. *Value distribution of holomorphic maps into compact, complex manifolds.* Lecture Notes in Mathematics. **135** (1970), pp. 267, Springer-Verlag.

[70] ———. *Value distribution of holomorphic maps.* Several Complex Variables I. Lecture Notes in Mathematics. **155** (1970), 165–190, Springer-Verlag.

[71] ———. *A Bezout estimate for complete intersections.* Ann. of Math. (2) **96** (1972), 361–401.

[72] ———. *Deficit and Bezout estimates.* Value Distribution Theory Part B (edited by R. O. Kujala and A. L. Vitter, III), Pure and Appl. Math. **25** Marcel Dekker, New York (1973), pp. 271.

[73] ———. *Holomorphic functions of finite order in several complex variables.* CBMS Regional Conference Series in Math. **21** Amer. Math. Soc. Providence, RI, (1974), pp. 83.

[74] ———. *Aspects of value distribution theory in several complex variables.* Bull. Amer. Math. Soc. **83** (1977), 166–183.

[75] ———. *Value distribution on parabolic spaces.* Lecture Notes in Mathematics **600** (1977), p. 216. Springer-Verlag.

[76] ———. *A Casorati-Weierstrass theorem for Schubert zeros of semi-ample, holomorphic vector bundles.* Atti. Acad. Naz. Lincei. Mem. C1, Sci. Fis. Mat. Natur. Ser. VIII m. **15** (1978), 63–90.

[77] ———. *The characterization of strictly parabolic manifolds.* Ann. Scuola Norm. Sup. Pisa, **7** (1980), 87–154.

[78] ———. *The characterization of strictly parabolic spaces.* Composite Mathematica, **44** (1981), 305–373.

[79] ———. *Introduction to value distribution theory of meromorphic maps.* Lecture Notes in Mathematics **950** (1983), 210–359. Springer-Verlag.

[80] ———. *The Ahlfors Weyl theory of meromorphic maps on parabolic manifolds.* Lecture Notes in Mathematics, **981** (1983), 101–219. Springer-Verlag.

[81] ______. *Value distribution and the lemma of the logarithmic derivative on polydiscs*. Intern. J. Math. Sci. **6** (1983), no. 4, 617–669.

[82] ______. *Value distribution theory for meromorphic maps*. Asp. Math. **E7** (1985), pp. 347. Vieweg.

[83] ______. *Algebroid reduction of Nevanlinna theory*. Complex Analysis III (C. A. Berenstein ed.). Lecture Notes in Mathematics **1277** (1987), 131–241. Springer-Verlag.

[84] ______. *On the propogation of dependences*. Pac. J. of Math. **139** (1989), 311–337.

[85] ______. *An extension of the theorem of Steinmetz-Nevanlinna to holomorphic curves*. Math. Ann. **282** (1988), 185–222.

[86] ______. *Value Distribution Theory in Several Complex Variables*. In preparation. To appear in China.

[87] Thie, P., *The Lelong number of a point of a complex analytic set*, Math. Ann. **172** (1967), 269–312.

[88] Tung, Ch. *The first main theorem on complex spaces*. (1973 Notre Dame Thesis pp. 320) Atti della Acc. Naz. d. Lincei. Serie VIII **15** (1979), 93–262.

[89] Vitter, A., *The lemma of the logarithmic derivative in several complex variables*, Duke Math. J. **44** (1977), 89–104.

[90] Weyl, H., and Weyl, J., *Meromorphic functions and analytic curves*. Annals of Math. Studies **12** Princeton Univ. Press, Princeton N.J. (1943) pp. 269.

[91] Wu, H., *Remarks on the first main theorem in equidistribution theory, I, II, III, IV*. J. Differential Geometry **2** (1968), 197–202, 369–384, ibid. **3** (1969), 83–94, 433–446.

[92] ______, *The equidistribution theory of holomorphic curves*. Annals of Math. Studies, **64** Princeton Univ. Press, Princeton, N.J. (1970) pp. 219.

[93] Wong, P. M., *Defect relations for maps on parabolic spaces and Kobayashi metrics on projective spaces omitting hyperplanes*, Thesis Notre Dame (1976), pp. 231.

[94] Wong, P. M. *On the Second Main Theorem of Nevanlinna Theory*. Amer. J. Math. **111** (1989), 549–583.

[95] ______. *On holomorphic curves in spaces of constant holomorphic sectional curvature*, preprint 1980, p. 20, to appear in Proc. Conf. in Compl. Geometry, Osaka, Japan (1991).

[96] Wong, P. M. and Ru, M. *Integral points in* $\mathbb{P}^n - \{2n+1$ *hyperplanes in general position*) Invent. Math. **106** (1991), 195–216.

THE NEVANLINNA ERROR TERM FOR COVERINGS GENERICALLY SURJECTIVE CASE

WILLIAM CHERRY

Nevanlinna theory [**Ne**] started as the theory of the value distribution of meromorphic functions. The so-called Second Main Theorem is a theorem relating how often a function is equal to a given value compared with how often, on average, it is close to that value. This theorem takes the form of an inequality relating the counting function and the mean proximity function by means of an error term. Historically, only the order of the error term was considered important, but motivated by Vojta's [**Vo**] dictionary between Nevanlinna theory and Diophantine approximations, Lang and others, see [**La**] and [**L-C**] for instance, have started to look more closely at the form of this error term.

Vojta has a number theoretic conjecture, analogous to the Second Main Theorem, where the absolute height of an algebraic point is bounded by an error term, which is independent of the degree of the point. This caused Lang to raise the question, "how does the degree of an analytic covering of **C** come into the error term in Nevanlinna theory?" The second part of [**L-C**] looks at Nevanlinna theory on coverings in order to answer this question. Noguchi [**No1**], [**No2**], and [**No3**] and Stoll [**St**] are among those who have previously looked at the Nevanlinna theory of coverings.

As part of Vojta's dictionary, the Nevanlinna characteristic function corresponds to the height of a rational point in projective space. For a number field F, there are two notions of height. There is a relative height and an absolute height. Given a point $P = (x_0, \ldots, x_n)$ in $\mathbf{P}^n(F)$, the *relative* height, $h_F(P)$ is defined by

$$h_F(P) = \sum_{v \in S} [F_v : \mathbf{Q}_v] \log \max_j |x_j|_v,$$

where S is the set of absolute values on F, and $[F_v : \mathbf{Q}_v]$ is the local degree. The *absolute* height $h(P)$ is the relative height divided by the global degree $[F : \mathbf{Q}]$ and is independent of the field F. The Nevanlinna characteristic function T_f, as defined in Part II of [**L-C**], corresponds to the relative height. As such, one wanted a second main theorem where the degree enters into the error term only as a factor multiplied by a universal expression independent of the degree. This is more or less what was achieved when $T_f(r)$ was larger than the degree, but when $T_f(r)$ was less than the degree, we could not get such a result, and it appeared that the error term depended on the degree in a more subtle way. However, this is to be expected because the classical second main theorem only holds when $T_f(r)$ is greater than one, and the condition that the *relative* T_f be greater than the degree is precisely the condition that the *absolute* T_f be greater than one. The main objective of this note is to show that when the Nevanlinna functions on coverings are normalized from the beginning by dividing by the degree, then the error term is independent of the degree, completely in line with Vojta's conjecture in the number theoretic case, and all the extraneous terms in [**L-C**] disappear.

Furthermore, by making two minor changes to the method in [**L-C**], following Griffiths-King [**G-K**], we are able to work with non-degenerate holomorphic maps from an analytic covering of $\mathbf{C}^m$ into an n-complex dimensional manifold, where $m \geq n$. This is more general than the equidimensional case treated in [**L-C**] and shows that the error term retains the same structure when the dimension of the domain space is larger than that of the range.

The main result of this note is the following Second Main Theorem:

Theorem. *Let $p: Y \to \mathbf{C}^m$ be a finite normal analytic covering of $\mathbf{C}^m$ which is unramified and non-singular above zero. Let X be an n-complex dimensional manifold, and let $f: Y \to X$ be a non-degenerate holomorphic map such that the "ramification" divisor R_f does not intersect Y<0>.*

Let:

$D = \sum_{j=1}^{q} D_j$ be a divisor with simple normal crossings of complexity k;

$L_j = L_{D_j}$ be the line bundle associated to D_j;

ρ_j be a hermitian metric on L_j;

Ω be a volume form on X;
κ be the metric on the canonical bundle associated to Ω;
η be a positive $(1,1)$ form on X such that $\eta^n/n! \geq \Omega$ and $\eta \geq c_1(\rho_j)$ for all j;
Assume that $f(y) \notin D$ for all $y \in Y$<0>.
Let:

$$B = \frac{b^{1/n}}{n}((q+1)q^{k/n} + \frac{1}{2}q^{2+k/n}\log 2);$$
$$b_1 = b_1(F_{\gamma_f^{1/n}}) \quad and \quad r_1 = r_1(F_{\gamma_f^{1/n}}),$$

where b is the constant of Lemma II.7.4 in **[L-C]**, *and depends only on Ω, D and η. Then, one has*

$$T_{f,\kappa}(r) + \sum_{j=1}^{q} T_{f,\rho_j}(r) - N_{f,D}(r) + N_{R_f}(r) - N_{p,\mathrm{Ram}}(r)$$
$$\leq \frac{n}{2}S(BT_{f,\eta}^{1+k/n}, \psi, b_1, r) - \frac{1}{2}\sum_{y \in <0>} \frac{\log \gamma_f(y)}{[Y : \mathbf{C}^m]} + 1,$$

for $r \geq r_1$ outside of a set of measure $\leq 2b_0(\psi)$.

Remarks. The symbols above, including the divisor R_f, which is the Griffiths-King ramification term, will be precisely defined in the sequel. Note that except for an additive term, which can be made to disappear by normalizing *f*, the Jacobian of f and the Jacobian of p at the points which are above zero, the error term is completely uniform in the functions p and f as well as in the degree of the covering. Also, the extraneous terms involving the degree which appear in **[L-C]** are not present here. Furthermore, when the error term function is expanded out, the constant which appears in front of the $\log T_{f,\eta}$ term is

$$\frac{n}{2}\left(1 + \frac{k}{n}\right)$$

which is better than the constant $n(n+1)$ appearing in Stoll **[St]**. The larger constants in Stoll result from his method of summing up projections onto Grassmannians via the "associated maps." By combining the equidimensional method used by Wong **[Wo]** and improved by Lang, with the ramification terms in Griffiths-King which depend on the choice of Jacobian section, rather than the Wronskian determinant which appears in Stoll, the error term obtained in the generically surjective case is identical to that of the equidimensional case, and, in particular, does not contain the unnecessary factors which arise from projective linear algebra.

For the proof of the above theorem, we follow Chapter IV of [**L-C**].

1. Preliminaries. Let $p : Y \to \mathbf{C}^m$ be a finite normal analytic covering of $\mathbf{C}^m$, and assume that Y is non-singular at the points above zero and that p is also unramified above zero.
Let:

$$[Y : \mathbf{C}^m] = \text{the degree of the covering;}$$
$$z = (z_1, \ldots, z_m) \text{ be the complex coordinates of } C^m;$$
$$\|z\|^2 = \sum_{j=1}^{m} z_j \bar{z}_j;$$
$$Y(r) = \{y \in Y : \|p(y)\| < r\};$$
$$Y[r] = \{y \in Y : \|p(y)\| \leq r\};$$
$$Y\langle r\rangle = \{y \in Y : \|p(y)\| = r\}.$$

Consider the following differential forms on $\mathbf{C}^m$:

$$\omega(z) = dd^c \log \|z\|^2;$$
$$\varphi(z) = dd^c \|z\|^2;$$
$$\sigma(z) = d^c \log \|z\|^2 \wedge \omega^{m-1}(z);$$
$$\Phi(z) = \prod_{j=1}^{m} \left(\frac{\sqrt{-1}}{2\pi} dz_j \wedge d\bar{z}_j \right).$$

The pullback of these forms to Y via p will be denoted by a subscript Y:

$$\omega_Y = p^*\omega, \quad \varphi_Y = p^*\varphi, \quad \sigma_Y = p^*\sigma, \quad \Phi_Y = p^*\Phi.$$

Note that σ_Y is closed and C^∞ away from $Y\langle 0\rangle$ and that

$$\int_{Y\langle r\rangle} \sigma_Y = [Y : \mathbf{C}^m].$$

The following form of the Green-Jensen integral formula will be needed. For a proof, see **[L-C]** Theorem IV.1.2.

Theorem 1 (Green-Jensen Formula). *Let α be a C^2 function from $Y \to \mathbf{C}$ except on a negligible set of singularities Z such that $Z \cap Y\langle 0\rangle = \emptyset$. Assume, in addition, that the following three conditions are satisfied:*

i) *$\alpha\sigma_Y$ is absolutely integrable on $Y\langle r\rangle$ for all $r > 0$.*
ii) *$d\alpha \wedge \sigma_Y$ is absolutely integrable on $Y[r]$ for all r.*
iii) $\lim\limits_{\varepsilon\to 0} \int\limits_{S(Z,\varepsilon)(r)} \alpha\sigma_Y = 0$ *for all r,*

where for sufficiently small ε, $S(Z,\varepsilon)(r)$ denotes the boundary of the tubular neighborhood of radius ε around the singularities $Z \cap Y[r]$, which is regular for all but a discrete set of values ε. Then

$$(A)\quad \int_0^r \frac{dt}{t} \int_{Y\langle t\rangle} d^c\alpha \wedge \omega_Y^{m-1} = \frac{1}{2}\int_{Y\langle r\rangle} \alpha\sigma_Y - \frac{1}{2}\sum_{y\in Y\langle 0\rangle} \alpha(y),$$

and

$$(B)\quad \int_0^r \frac{dt}{t} \int_{Y(t)} dd^c\alpha \wedge \omega_Y^{m-1} + \int_0^r \frac{dt}{t} \lim_{\varepsilon\to 0} \int_{S(Z,\varepsilon)(t)} d^c\alpha \wedge \omega_Y^{m-1}$$
$$= \frac{1}{2}\int_{Y\langle r\rangle} \alpha\sigma_Y - \frac{1}{2}\sum_{y\in Y\langle 0\rangle} \alpha(y).$$

Let $f : Y \to X$ be a non-degenerate (i.e. not contained in any divisor on X) holomorphic map, where X is a compact n-complex dimensional manifold and n is assumed less than or equal to m.

Remark. It is not necessary for the function f to be defined on all of Y. Everything in the sequel remains true for a function $f : Y(R) \to X$ provided that $r < R$.

We define the **absolute** Nevanlinna functions as follows:

Height

If η is a $(1,1)$ form on X, then define

$$T_{f,\eta}(r) = \frac{1}{[Y : \mathbf{C}^m]} \int_0^r \frac{dt}{t} \int_{Y(t)} f^*\eta \wedge \omega_Y^{m-1},$$

and similarly, given a hermitian metric ρ on a holomorphic line bundle L on X, define

$$T_{f,\rho}(r) = \frac{1}{[Y : \mathbf{C}^m]} \int_0^r \frac{dt}{t} \int_{Y(t)} f^*c_1(\rho) \wedge \omega_Y^{m-1},$$

where $c_1(\rho) = dd^c \log \rho$ is the **Chern form** of ρ.

Counting functions

Given a divisor D on Y, let

$$\mathbf{n}_D(t) = \frac{1}{[Y : \mathbf{C}^m]} \int_{D(t)} \omega_Y^{m-1} \quad \text{and} \quad N_D(r) = \int_0^r \mathbf{n}_D(t) \frac{dt}{t},$$

and given a divisor D on X, let $N_{f,D} = N_{f^*D}$. The counting function for the ramification divisor of p, defined locally by the zeros of the Jacobian matrix, will be denoted $N_{p,\mathrm{Ram}}(r)$.

A volume form Ω on X defines a metric κ on the canonical line bundle K of X. Since,

$$f^*\mathrm{Ric}\,\Omega = f^*c_1(\kappa),$$

the **height associated to the volume form** Ω is defined as:

$$T_{f,\kappa}(r) = \frac{1}{[Y : \mathbf{C}^m]} \int_0^r \frac{dt}{t} \int_{Y(t)} f^* c_1(\kappa) \wedge \omega_Y^{m-1}.$$

2. Ramification. Let Φ be the Euclidean volume form on $\mathbf{C}^m$ and let $\Phi_Y = p^*(\Phi)$ be the pullback to a pseudo-volume form on Y. Let Ω be a volume form on X. Because f is non-degenerate, we can assume that the coordinates on $\mathbf{C}^m$ were chosen so that

$$f^*\Omega \wedge p^*\left(\prod_{j=n+1}^{m} \frac{\sqrt{-1}}{2\pi} dz_j \wedge d\bar{z}_j\right)$$

is not identically zero. Following Griffiths and King [**G-K**], let γ_f be the non-negative function such that

$$f^*\Omega \wedge p^*\left(\prod_{j=n+1}^{m} \frac{\sqrt{-1}}{2\pi} dz_j \wedge d\bar{z}_j\right) = \gamma_f \Phi_Y.$$

Note that γ_f is singular along the ramification divisor of p and vanishes along the divisor R_f given by the equation

$$f^*\Omega \wedge p^*\left(\prod_{j=n+1}^{m} \frac{\sqrt{-1}}{2\pi} dz_j \wedge d\bar{z}_j\right) = 0.$$

Remark. When $n = m$, the divisor R_f is the ramification divisor associated to the map f. In general, the divisor R_f depends not only on the ramification of f, but also on the choice of coordinates on $\mathbf{C}^m$. However, this dependence on the choice of coordinates is omitted from the notation.

Note that because $f^*c_1(\kappa) = dd^c \log \gamma_f$, one has

$$T_{f,\kappa}(r) = \frac{1}{[Y : \mathbf{C}^m]} \int_0^r \frac{dt}{t} \int_{Y(t)} dd^c \log \gamma_f \wedge \omega_Y^{m-1}.$$

Theorem 2. *Assume that $p : Y \to \mathbf{C}^m$ is unramified above zero, and let $f : Y \to X$ be a non-degenerate holomorphic map such that the divisor R_f does not intersect $Y\langle 0\rangle$. Then*

$$T_{f,\kappa}(r) + N_{R_f}(r) - N_{p,\mathrm{Ram}}(r)$$
$$= \frac{1}{2} \int\limits_{Y\langle r\rangle} (\log \gamma_f) \frac{\sigma_Y}{[Y : \mathbf{C}^m]} - \frac{1}{2} \sum_{y \in Y\langle 0\rangle} \frac{\log \gamma_f(y)}{[Y : \mathbf{C}^m]}.$$

Proof: With new notation, this is simply Theorem 1 (B) combined with the fact that

$$\lim_{\varepsilon \to 0} \int\limits_{S(Z,\varepsilon)(t)} d^c \log \gamma_f \wedge \omega_Y^{m-1} = \int\limits_{Z(t)} \omega_Y^{m-1},$$

where Z is the set of singularities for $\log \gamma_f$, and then divided by the degree.

3. Calculus Lemmas. Let ψ be a positive increasing function, such that

$$\int\limits_e^\infty \frac{du}{u\psi(u)} = b_0(\psi)$$

is finite. Such a function is called a **type function.** Given a positive increasing function F, let $r_1(F)$ be the smallest number such that $F(r) \geq e$ for $r \geq r_1(F)$, and let $b_1(F)$ be the smallest number greater than or equal to one, such that

$$b_1 r^{2m-1} F'(r) \geq e \quad \text{for all} \quad r \geq r_1(F).$$

Define the **error term** function to be

$$S(F, b_1, \psi, r) = \log \{F(r)\psi(F(r))\psi(r^{2m-1} b_1 F(r)\psi(F(r)))\}.$$

Given a function α on Y, define the **height transform**:

$$F_\alpha(r) = \frac{1}{[Y:\mathbf{C}^m]} \int_0^r \frac{dt}{t^{2m-1}} \int_{Y(t)} \alpha \Phi_Y$$

for $r > 0$.

Let α be a function on Y such that the following conditions are satisfied:

(a) α is continuous and > 0 except on a divisor of Y.

(b) For each r, the integral $\int_{Y<r>} \alpha\sigma_Y$ is absolutely convergent and $r \mapsto \int_{Y<r>} \alpha\sigma_Y$ is a piecewise continuous function of r.

(c) There is an $r_1 \geq 1$ such that $F_\alpha(r_1) \geq e$.

Note: F_α has positive derivative, so is strictly increasing.

Lemma 3. *If* α *satisfies* (a), (b) *and* (c) *above, then* F_α *is* C^2 *and*

$$\frac{1}{r^{2m-1}} \frac{d}{dr}(r^{2m-1} F'_\alpha(r)) = \frac{2}{(m-1)!} \int_{Y<r>} \alpha \frac{\sigma_Y}{[Y:\mathbf{C}^m]}.$$

Proof: Use Fubini's Theorem and the fact that

$$\Phi_Y = \frac{\|p\|^{2(m-1)}}{(m-1)!} d\|p\|^2 \wedge \sigma_Y,$$

as in Chapter IV §3, and then divide by the degree.

The standard Nevanlinna calculus lemma then gives

Lemma 4. *If* α *satisfies* (a), (b) *and* (c) *above, then*

$$\log \int_{Y<r>} \alpha \frac{\sigma_Y}{[Y:\mathbf{C}^m]} \leq S(F_\alpha, b_1(F_\alpha), \psi, r) + \log \frac{(m-1)!}{2}$$

for all $r \geq r_1(F_\alpha)$ *outside a set of measure* $\leq 2b_0(\psi)$.

4. Trace and Determinant. Given a $(1,1)$ form η on Y, define the **trace** and **determinant** outside the ramification points of p as follows:

$$(\det(\eta))\Phi_Y = \frac{1}{m!}\eta^m$$
$$(m-1)!\,\mathrm{tr}(\eta)\Phi_Y = \eta \wedge \varphi_Y^{m-1}.$$

Furthermore, define the **n × n trace** and **determinant** outside the ramification points of p as follows:

$$(\det_n(\eta))\Phi_Y = \frac{1}{n!}\eta^n \wedge p^*\left(\prod_{j=n+1}^{m} \frac{\sqrt{-1}}{2\pi} dz_j \wedge d\bar{z}_j\right)$$
$$(n-1)!\,\mathrm{tr}_n(\eta)\Phi_Y = \eta \wedge \varphi_Y^{n-1} \wedge p^*\left(\prod_{j=n+1}^{m} \frac{\sqrt{-1}}{2\pi} dz_j \wedge d\bar{z}_j\right).$$

In the case when Y is $\mathbf{C}^m$ and p is the identity, the $n \times n$ trace and determinant are simply the trace and determinant of the $n \times n$ block in the upper-left of the matrix corresponding to η.

The following lemma is simply the pull-back to Y of some relations on $\mathbf{C}^m$, which follow immediately from the elementary linear algebra of hermitian positive semi-definite matrices.

Lemma 5. *If η is a semi-positive $(1,1)$ form on Y, then*

$$(\det_n(\eta))^{1/n} \leq \frac{1}{n}\,\mathrm{tr}_n(\eta) \quad \textit{and} \quad \mathrm{tr}_n(\eta) \leq \mathrm{tr}\,(\eta)$$

for the regular points in Y which are not ramification points of p.

Let η be a closed, positive $(1,1)$ form such that

$$\Omega = \frac{1}{n!}\eta^n.$$

Since

$$f^*\Omega \wedge p^*\left(\prod_{j=n+1}^{m} \frac{\sqrt{-1}}{2\pi} dz_j \wedge d\bar{z}_j\right) = \gamma_f \Phi_Y,$$

one finds that $\gamma_f = \det_n(f^*\eta)$.

Proposition 6. *Let* $\tau_f = \mathrm{tr}_n(f^*\eta)$. *Then*

$$(m-1)!F_{\tau_f} \leq T_{f,n}.$$

Proof: Let $\tau'_f = \mathrm{tr}\,(f^*\eta)$. All the symbols have been defined so that the proof of Proposition II.6.2 in [**L-C**], after dividing through by the degree, gives

$$T_{f,\eta} = (m-1)!F_{\tau'_f}.$$

But, since $\mathrm{tr}_n(f^*\eta) \leq \mathrm{tr}\,(f^*\eta)$, one has $F_{\tau_f} \leq F_{\tau'_f}$.

5. Second Main Theorem. Replacing the counterparts to the statements above in the proof of Theorem IV.4.3 in [**L-C**] gives the following Second Main Theorem.

Theorem 7. *Assume that* $p : Y \to \mathbf{C}^m$ *is unramified above zero, and let* $f : Y \to X$ *be a non-degenerate holomorphic map such that the divisor* R_f *does not intersect* $Y\langle 0\rangle$. *Let* $T_{f,\kappa}$ *be the height associated to the volume form* $\Omega = \eta^n/n!$ *on* X. *Then*

$$T_{f,\kappa}(r) + N_{R_f}(r) - N_{p,\mathrm{Ram}}(r) \leq \frac{n}{2}S(T_{f,\eta}, b_1(F_{\tau_f}), \psi, r) - \frac{1}{2}\sum_{y\in Y\langle 0\rangle} \frac{\log \gamma_f(y)}{[Y:\mathbf{C}^m]}$$

for all $r \geq r_1(F_{\tau_f})$ *outside a set of measure* $\leq 2b_0(\psi)$.

Proof:

$$T_{f,\kappa}(r) + N_{f,\mathrm{Ram}}(r) - N_{p,\mathrm{Ram}}(r) + \frac{1}{2}\sum_{y\in Y\langle 0\rangle} \frac{\log \gamma_f(y)}{[Y:\mathbf{C}^m]}$$

$$= \frac{1}{2}\int_{Y\langle r\rangle} (\log \gamma_f)\frac{\sigma_Y}{[Y:\mathbf{C}^m]} \qquad \text{[Theorem 2]}$$

$$= \frac{n}{2}\int_{Y\langle r\rangle} \log \gamma_f^{1/n}\frac{\sigma_Y}{[Y:\mathbf{C}^m]}$$

$$\leq \frac{n}{2} \log \int_{Y<r>} \gamma_f^{1/n} \frac{\sigma_Y}{[Y : \mathbf{C}^m]}$$

[concavity of the log]

$$\leq \frac{n}{2} \log \int_{Y<r>} \tau_f \frac{\sigma_Y}{[Y : \mathbf{C}^m]}$$

[Lemma 5]

$$\leq \frac{n}{2} S(F_{\tau_f}, b_1(F_{\tau_f}), \psi, r) + \frac{n}{2} \log(m-1)!$$

[Lemma 4]

$$\leq \frac{n}{2} S(T_{f,\eta}, b_1(F_{\tau_f}), \psi, r)$$

[Proposition 6]

for all $r \geq r_1(F_{\tau_f})$ outside a set of measure $\leq 2b_0(\psi)$.

Remarks. The term on the right involving $\log \gamma_f$ in the above inequality depends only on the values of *f*, the Jacobian of *f*, and the Jacobian of p above zero, so if these functions are normalized at the points above zero, then the right hand side is uniform in the functions f and p and in the degree. In the case when $m = n$, then $F_{\tau_f} = T_{f,\eta}/(n-1)!$, so $r_1(F_{\tau_f}) = r_1(T_{f,\eta}/(n-1)!)$.

Similar changes give the more general Second Main Theorem. Recall that a divisor is said to have simple normal crossings of complexity k if k is the minimal number such that there exist local coordinates $w_1, \ldots, w_n$ around each point of D, such that D is defined locally by $w_1 \ldots w_l = 0$, with $l \leq k$.

For the rest of this section, let:

$D = \sum_{j=1}^{q} D_j$ be a divisor on X with simple normal crossings of complexity k;

$L_j = L_{D_j}$ the holomorphic line bundle associated to D_j with hermitian metric ρ_j;

η be a closed, positive $(1,1)$ form on X such that $\eta \geq c_1(\rho_j)$ for all j, and $\eta^n/n! \geq \Omega$;

s_j be a holomorphic section of L_j, such that $(s_j) = D_j$;

Since X is compact, after possibly multiplying s_j by a constant, assume without loss of generality that

$$|s_j|_{\rho_j} \leq 1/e \leq 1.$$

For convenience, also assume that $f(y) \notin D$ for all $y \in Y<0>$, and that $Y<0>$ does not intersect the ramification divisor of f.

If λ is a constant with $0 < \lambda < 1$, then define the **Ahlfors-Wong** singular volume form

$$\Omega(D)_\lambda = \Big(\prod |s_j|_j^{-2(1-\lambda)}\Big)\Omega,$$

and define

$$\gamma_\lambda = \prod |s_j \circ f|_j^{-2(1-\lambda)} \gamma_f.$$

Given Λ a positive decreasing function of r with $0 < \Lambda < 1$, define

$$\gamma_\Lambda = \prod |s_j \circ f|_j^{-2(1-\Lambda)} \gamma_f.$$

Note that because of the assumption $|s_j|_j \leq 1/e \leq 1$, one has $\gamma_f \leq \gamma_\Lambda$.

Using the fact that $\mathrm{tr}_n \leq \mathrm{tr}$ and dividing through by the degree in the proof of Lemma IV.5.1 in **[L-C]** gives the following lemma.

Lemma 8. *Let b be the constant of Lemma II.7.4 in* **[L-C]**, *which depends only on Ω, D and η. Then for any decreasing function Λ with $0 < \Lambda < 1$, one has*

$$F_{\gamma_\Lambda^{1/n}}(r) \leq (q+1)\frac{b^{1/n}}{n(m-1)!}\frac{T_{f,\eta}(r)}{(\Lambda(r))^{k/n}} + \frac{qb^{1/n}}{2n(m-1)!}\frac{\log 2}{(\Lambda(r))^{1+k/n}}$$

for all r.

Remark. Notice that the degree no longer appears in this estimate, and this is why the error term is now uniform. Also note that the $n!$ in the denominator has been replaced with $n(m-1)!$.

Let $r_1 = r_1(F_{\gamma_f^{1/n}})$ and let

$$\Lambda(r) = \begin{cases} \dfrac{1}{qT_{f,\eta}(r)} & \text{for } r \geq r_1 \\ \text{constant} & \text{for } r \leq r_1. \end{cases}$$

Note that since $\eta^n/n! \geq \Omega$, one has $F_{\gamma_f^{1/n}} \leq T_{f,\eta}/n!$. Therefore $r_1(F_{\gamma_f^{1/n}}) \geq r_1(T_{f,\eta}/n!)$, and hence one has $\Lambda \leq 1$.

Applying Lemma 8 to the function Λ proves the next lemma.

Lemma 9. *Let b be the constant of Lemma II.7.4 of* [**L-C**] *and let*

$$B = \frac{b^{1/n}}{n}((q+1)q^{k/n} + \frac{1}{2}q^{2+k/n}\log 2).$$

Then

$$F_{\gamma_\Lambda^{1/n}}(r) \leq \frac{B}{(m-1)!}T_{f,\eta}^{1+k/\eta}$$

for $r \geq r_1$.

Lemma 10. *One has*

$$\log \int_{Y\langle r\rangle} \gamma_\Lambda^{1/n} \frac{\sigma_Y}{[Y : \mathbf{C}^m]} \leq S(BT_{f,\eta}^{1+k/n}, b_1, \psi, r)$$

for all $r \geq r_1$, outside a set of measure $\leq 2b_0(\psi)$, where

$$B = \frac{b^{1/n}}{n}((q+1)q^{k/n} + \frac{1}{2}q^{2+k/n}\log 2)$$
$$b_1 = b_1(F_{\gamma_f^{1/n}}) \quad \textit{and} \quad r_1 = r_1(F_{\gamma_f^{1/n}}).$$

Proof: Because $\gamma_\Lambda \geq \gamma_f$, one has

$$F_{\gamma_\Lambda^{1/n}} \geq F_{\gamma_f^{1/n}} \quad \text{and} \quad F'_{\gamma_\Lambda^{1/n}} \geq F'_{\gamma_f^{1/n}}.$$

Hence $b_1 = b_1(F_{\gamma_f^{1/n}})$ and $r_1 = r_1(F_{\gamma_f^{1/n}})$ are such that for $r \geq r_1$,

$$F_{\gamma_\Lambda^{1/n}}(r) \geq e \quad \text{and} \quad b_1 r^{2n-1} F'_{\gamma_\Lambda^{1/n}}(r) \geq e.$$

From Lemma 4, one has

$$\log \int_{Y\langle r\rangle} \gamma_\Lambda^{1/n} \frac{\sigma_Y}{[Y : \mathbf{C}^m]} \leq S(F_{\gamma_\Lambda^{1/n}}, b_1, \psi, r) + \log \frac{(m-1)!}{2}$$

for $r \geq r_1$ outside an exceptional set of measure $\leq 2b_0(\psi)$. Now from Lemma 9, one has

$$S(F_{\gamma_\Lambda^{1/n}}, b_1, \psi, r) + \log \frac{(m-1)!}{2} \leq S(BT_{f,\eta}^{1+k/n}, b_1, \psi, r)$$

for $r \geq r_1$.

Finally, we can prove the general Second Main Theorem.

Theorem 11. *One has*

$$\begin{aligned} T_{f,\kappa}(r) &+ \sum_{j=1}^{q} T_{f,\rho_j}(r) - N_{f,D}(r) + N_{R_f}(r) - N_{p,\mathrm{Ram}}(r) \\ &\leq \frac{n}{2} S(BT_{f,\eta}^{1+k/n}, \psi, b_1, r) - \frac{1}{2} \sum_{y \in Y\langle 0\rangle} \frac{\log \gamma_f(y)}{[Y : \mathbf{C}^m]} + 1, \end{aligned}$$

for $r \geq r_1$ outside of a set of measure $\leq 2b_0(\psi)$.

Proof: Let λ be a constant with $0 < \lambda < 1$. Using Theorem 1 (B), and the fact that $dd^c \log$ transforms products into sums, one obtains:

$$\begin{aligned} T_{f,\kappa}(r) + (1-\lambda) \sum_{j=1}^{q} T_{f,\rho_j}(r) &- (1-\lambda) \sum_{j=1}^{q} N_{f,D_j}(r) \\ &+ N_{R_f}(r) - N_{p,\mathrm{Ram}}(r) \end{aligned}$$

$$= \frac{1}{[Y:\mathbf{C}^m]} \int_0^r \frac{dt}{t} \int_{Y(t)} dd^c \log \gamma_\lambda + \frac{1}{[Y:\mathbf{C}^m]} \int_0^r \frac{dt}{t} \lim_{\epsilon \to 0} \int_{S(Z,\epsilon)(t)} d^c \log \gamma_\lambda$$

$$= \frac{n}{2} \int_{Y<r>} \log \gamma_\lambda^{1/n} \frac{\sigma_Y}{[Y:\mathbf{C}^m]} - \frac{1}{2} \sum_{y \in Y<0>} \frac{\log \gamma_\lambda(y)}{[Y:\mathbf{C}^m]}.$$

Because of the assumption that $|s_j|_j \leq 1$, one also has

$$-\frac{1}{2} \sum_{y \in Y<0>} \log \gamma_\lambda(y) \leq -\frac{1}{2} \sum_{y \in Y<0>} \log \gamma_f(y).$$

Also, since Λ is constant on $Y<r>$, the function Λ can replace λ in the above equality. Furthermore, $N_{f,D_j} \geq 0$ and $-1 \leq -(1-\lambda)$, so the factor $(1-\lambda)$ in front can be deleted. When $r \geq r_1$, one has

$$-1 \leq -\Lambda(r) \sum_{j=1}^{q} T_{f,\rho_j}(r),$$

from the definition of Λ, and from the fact that η was chosen so that

$$T_f, \eta \geq T_{f,\rho_j} \text{ for all } j.$$

Finally, by moving the log out of the integral, one has

$$\int_{Y<r>} \log \gamma_{\Lambda(r)}^{1/n} \frac{\sigma_Y}{[Y:\mathbf{C}^m]} \leq \log \left(\int_{Y<r>} \gamma_{\Lambda(r)}^{1/n} \frac{\sigma_Y}{[Y:\mathbf{C}^m]} \right).$$

Applying the estimate in Lemma 10 for Λ to the term with the integral on the right and collecting terms concludes the proof of the theorem.

References

[G-K] P. Griffiths and J. King, *Nevanlinna theory and holomorphic mappings between algebraic varieties*, Acta Mathematica **130** (1973), 145–220.

[La] S. Lang, *The error term in Nevanlinna theory* II, Bull. AMS (1990) pp. 115–125.

[L-C] S. Lang and W. Cherry, *Topics in Nevalinna Theory*, Lecture Notes in Mathematics 1433, Springer Verlag, 1990.

[Ne] R. Nevanlinna, *Analytic Functions*, Springer Verlag, 1970; (revised translation of the German edition, 1953).

[No1] J. Noguchi, *A relation between order and defects of meromorphic mappings of* $\mathbf{C}^n$ *into* $\mathbf{P}^N(\mathbf{C})$, Nagoya Math. J. **59** (1975), 97–106.

[No2] J. Noguchi, *Meromorphic mappings of a covering space over* $\mathbf{C}^m$ *into a projective variety and defect relations*, Hiroshima Math. J. **6** (1976), 265–280.

[No3] J. Noguchi, *On the value distribution of meromorphic mappings of covering spaces over* $\mathbf{C}^m$ *into algebraic varieties*, J. Math. Soc. Japan. **37** (1985), 295–313.

[St] W. Stoll, *The Ahlfors-Weyl theory of meromophic maps on parabolic manifolds*, Lecture Notes in Mathematics 981, Springer Verlag, 1981.

[Vo] P. Vojta, *Diophantine Approximations and Value Distribution Theory*, Lecture Notes in Mathematics 1239, Springer Verlag, 1987.

[Wo] P. M. Wong, *On the second main theorem of Nevanlinna theory*, Am. J. Math. **111** (1989), pp. 549–583.

ON AHLFORS'S THEORY OF COVERING SURFACES

DAVID DRASIN

To Wilhelm Stoll, The Vincent F. Duncan and
Annamarie Micus Duncan Professor of Mathematics

1. Introduction. In [1] (see [3, pp. 214-251]) Ahlfors introduced his theory of covering surfaces. His approach was combinatorial and geometric, and showed that R. Nevanlinna's theory of meromorphic functions had topological significance, and held in differentiated form. Other accounts are in [2], [5], [9], and [12] presents a very efficient proof using Ahlfors's own framework. See [10] for an independent approach, where the conclusions are slightly weaker than in [1].

Some years ago, John Lewis asked me if there was a way to derive Nevanlinna's value distribution theory directly from the argument principle. Since Nevanlinna's approach is based on Jensen's formula, itself the integrated argument principle, it is clear that the argument principle lies behind the theory, but the connection is, to say the least, highly indirect.

In this paper we show a more transparent connection. Very little is used that is not in a first course in complex analysis, but the subtleties needed to achieve (1.5) and (1.6) show the depth of Ahlfors's own insights. In retrospect our methods have considerable intersection with those of [1], although the orientation is different. I thank H. Donnelly, A. Eremenko, D. Gottlieb, L. Lempert, M. Ramachandran and A. Weitsman for helpful discussions. The idea for the latter part of Proposition (1.8) was shown to me by S. Lalley. The influence of Miles's work [8] is also apparent; see (2.23) below.

(1.1) Preliminaries. (See [1], [9, Ch. 13].) Let $a_1, \dots, a_q$ be distinct (finite) complex numbers. We develop two situations in parallel: the "base surface" F_0 is either the Riemann sphere S^2 or $S^2 \setminus \bigcup_{k=1}^{q} D_k$, where the D_k are disjoint continuua about the a_k; we also let $a_{q+1} = \infty$ and take D_{q+1} accordingly. Thus, F_0 is either closed or bordered.

We impose a unit mass $\lambda(w)$ on F_0 with the properties specified in [1, I.1], [9, p. 325]; this allows lengths to be assigned to (Ahlfors) regular curves and open sets. The essential property of λ is that an isoperimetric inequality hold locally: each point $p_0 \in F_0$ has a neighborhood $U = U(p)$ such that if γ is a simple closed curve in U which bounds the region $\Omega \subset U$, then

$$\lambda(\Omega) < h\lambda(\gamma), \tag{1.2}$$

where $h = h(p)$. By compactness (1.2) holds on F_0 with a universal h so long as the λ-area of Ω is strictly bounded from one. We also let $[\cdot]$ be the chordal metric on S^2; it clearly satisfies (1.2).

Let $\Delta(r) = \{|z| < r\}$, $B(r) = \partial\Delta(r)$, $\Delta_\lambda(w_0, \eta) = \{w; \lambda(w, w_0) < \eta\}$, $B_\lambda(w_0, \eta) = \partial\Delta_\lambda(w_0, \eta)$, and $\Gamma_r = f(B(r))$. We consider maps $f : \Delta(R) \to S^2$ which preserve orientation, with $0 < R \leq \infty$. The most important setting is that f be meromorphic, but it is natural to require only that f be a ramified covering of S^2, in the spirit of Ahlfors-Sario [4]; according to Stoïlow [11] such maps become meromorphic if $\Delta(R)$ is given an appropriate structure. For a good account of this see, for example, [6, §2].

Consider now $f^{-1}(F_0) \subset \Delta(R)$; this is a union of components $\{G\}$. Let G be one such component. Then for $0 < r < R$, set $G(r) = G \cap \Delta(r)$, $\partial G(r) = G \cap B(r)$, and define in terms of λ the expressions $S = S(r)$, $L = L(r)$ for the area (including multiplicity) of $f(G(r))$ and length of $\Gamma_r = f(\partial G(r))$, measured by λ. In this sense, the image $\mathfrak{F}$ of G by f is a *covering surface* over F_0 with $\pi : \mathfrak{F} \to F_0$ the projection. Ahlfors also considers any λ-measurable subset $D \subset F_0$, and defines

$$S(r, D) = \frac{\lambda((\pi^{-1}(D) \cap G)(r))}{\lambda(D)}.$$

For example, if f is rational of degree N and D is any open set then $S(r) = N + o(1) = S(r, D)$ and $L(r) = o(1)(r \to \infty)$.

Ahlfors's theory has significance primarily when $\mathfrak{F}$ is *regularly exhaustible:* there exists an r-set A, R a limit point of A, such that

$$L(r) = o(S(r)) \quad (r \to R, r \in A) \tag{1.3}$$

[5, p. 338]. If f is meromorphic or quasiregular (and nonconstant) in $\Delta(\infty)$ and $\lambda = [\,]$, it requires but a few lines and the Schwarz inequality to see that the full image $\mathfrak{F}$ of f over $F_0 = S^2$ is regularly exhaustible (cf. [5, p. 352] and [1, p. 186]). In particular, in this case A consists of nearly all large r.

We state Ahlfors's conclusions in two forms:

(1.4) THEOREM. *(A) Let f be meromorphic in $\Delta(R)$, and $a_1, \ldots, a_q$ be distinct (finite) complex numbers. Then there exists $h = h(a_1, \ldots, a_q) > 0$ such that*

$$\sum_{1}^{q} n(r, a_j) > (q-2)S(r) - hL(r). \tag{1.5}$$

(B) Let F_0 be S^2 or $S^2 \setminus \cup D_k$, $f : \Delta(R) \to F_0$, and let λ be a unit mass as described above. Then there exists $h = h(F_0) > 0$ such that the Euler characteristic of any (finite) covering surface $\mathfrak{F}$ over F_0 satisfies

$$\rho^+ \equiv \max(\rho, 0) \geq \rho_0 S - hL. \tag{1.6}$$

Remarks. 1. In (1.6) we use the definitions of ρ and ρ_0 from [1]; cf. [4, p. 55]: $\chi = -F + E - V$ (F = faces, E = edges, V = vertices). In many contemporary topological texts, what we call ρ is considered the negative of the Euler characteristic.

2. Following [9], we assume that F_0 is planar: $S^2 \setminus \cup D_j$. Ahlfors observes [1, p. 174] that the general case follows from this by an elementary combinatorial analysis.

3. Inequality (1.6) is formally stronger than (1.5). One way to see this is that when (1.6) is used to derive the differentiated Nevanlinna theory (cf. [5, p. 148]) there is an additional branching term that is not apparent in (1.5). However, the arguments used to get these refinements are somewhat intricate; here we find that a common attack can yield both.

4. In accord with standard tradition, we use h as a positive constant which can be taken to depend only on data of the surface F_0.

For example, Picard's theorem is an immediate consequence of (1.5) or (1.6) together with (1.3); we consider (1.6). Let f be nonconstant on $\Delta(\infty)$ and omit a_1, a_2, a_3. Let F_0 be S^2 with small disks D_k deleted about the a_k, so that $\rho_0 = 1$. By assumption, no inverse image of any D_k can be compactly contained in any $\Delta(r)$, so that always $\rho \equiv -1$. Thus (1.3) and (1.5) are incompatible. Miles [7] shows that the main part of Nevanlinna's second fundamental theorem can be recovered from the Ahlfors theory.

(1.7) Normal values, first fundamental theorem. Since the argument principle gives $n(r, a_k) - n(r, \infty)$ rather than $n(r, a_k)$ directly, we first show that f always has many "normal" values. We have

(1.8) PROPOSITION. *Let $r < R$, $w_0 \in S^2$ and $\eta_0 > 0$ be given. Then there exist $K < \infty$ and $w^* \in S^2$, with $[w^*, w] < \eta_0$ and w^* normal in the sense*

$$|n(r, w^*) - S(r)| < KL(r);$$

further, we may find a line L through w^* such that Γ_r intersects $L \cap \{[w, w^*] < \eta_0\}$ in at most $KL(r)$ points.

Proof. We assume the elementary first covering theorem of Ahlfors (the analogue of Nevanlinna's first fundamental theorem; cf. [9, pp. 328-9]): if D is an open set in F_0 then $|S(r) - S(r, D)| < h(\lambda(D))^{-1}L(r)$. [Ahlfors also has a variant of this for coverings of "regular" curves, but that is not needed here].

In this proof, we take λ to be chordal measure [] on S^2, and let $S(r)$, $S(r, D)$ be computed with respect to [].

For a fixed (large) K, let $D_1 = \{w \in F_0; n(r, w) > S(r) + KL(r)\}$, and $D_2 = \{w; n(r, w) > S(r) - KL(r)\}$; here $n(r, w)$ is the usual counting function of w-values in $\Delta(r)$ or $G(r)$. Since $S(r, D_1) = (\int_{D_1} n(r, w) d\lambda(w))\{\lambda(D_1)\}^{-1}$, the first covering theorem yields that $KL(r)\lambda(D_1) < hL(r)$; thus if K is large, $\lambda(D_1)$ is bounded away from 1. The same analysis applies to D_2, and hence if K is sufficiently large, the set W of w^* which satisfy the Proposition has chordal measure at least .9 the measure of the ball $\{[w, w_0] < \eta_0\}$.

To satisfy the second condition, let us assume that $w_0 = 0$ and, since η_0 is small, replace the chordal metric by the Euclidean metric. Write $\Gamma = \Gamma_r$, and assume $\lambda(\Gamma) < \infty$. By making a rotation, we may assume that the intersection of Γ with each horizontal or vertical line contains no segment. We will show that if $i(y_0)$ is the cardinality of $\Gamma_r \cap \{\Im z = y_0\} \cap \{|z| < 1\}$, then there exists a set Y of y, $-\frac{1}{2}\eta_0 < y < \frac{1}{2}\eta_0$ with $\int_Y dt > .9\eta_0$ and

$$(1.9) \qquad i(y) < KL(r), \qquad y \in Y,$$

$(K = K(\eta))$.

If we grant this, it follows that there exists $y_0 \in Y, |y_0| < \frac{1}{4}\eta_0$, such that the set $\{y = y_0\} \cap \{|w| < \frac{1}{2}\eta_0\}$ has nonempty intersection with the set W constructed above. We use any $w^* = x_0 + iy_0$ in W, with $|x_0| < \frac{1}{2}\eta, y_0 \in Y$, and see that it satisfies both conditions of the Proposition.

We now produce y_0 so that (1.9) holds. By our normalization, $\Gamma \cap \Delta(w_0, 1)$ may be written as an at most countable union of graphs of continuous functions, say $y = y_j(x)$, $\alpha_j \le x \le \beta_j$, with $-1 < y_j(x) < 1$. If V_j is the total variation of y_j on (α_j, β_j) and L_j is the length of the graph of y_j, we have that $V_j \le L_j$.

Let $i_j(y)$ be the number of points of intersection of the graph of y_j with the line $\{\Im z = y\}$, so that $i(y) = \sum_j i_j(y)$; then Banach's formula for total variation gives that $V_j = \sum_{-1}^{1} i_j(y)dy$. Hence, $\int_{-1}^{1} i(y)dy < L(r)$, so that (1.9) follows at once.

(1.10) NORMALIZATION. *Given a fixed r, we in general take $w^* = \infty$ in Proposition 1.8.*

2. Partitioning of $\Delta(\mathbf{r})$. Given distinct complex numbers $a_1, \ldots, a_q$, let $10^{10}\eta < \inf_{i \ne j} \lambda(a_i, a_j)$. By Proposition 1.8, we may, by decreasing η if necessary, choose a_{q+1} so that $\lambda(a_{q+1}, a_k) > 10^{10}\eta (1 \le k \le q)$ and then, after a Möbius transformation of f assume that $a_{q+1} = \infty$. This choice of η is in force for all that follows, so that η depends only on F_0. Following the ideas of Ahlfors, construct (indexing mod $q+1$) Jordan arcs $\beta_k (1 \le k \le q+1)$ to join a_k to a_{k+1}. The β's divide F_0 with two Jordan domains F' and F'', and the preimages of the β's divide $\Delta(r)$ (or G, as appropriate) into N domains G_α. We let $F_\alpha = f(G_\alpha)$, so that F_α is contained in F' or F''. We usually ignore the specific choice of F' or F'', and write that $f(G_\alpha) \subset F$, where F is the relevant choice of F' or F''.

Depending on the context, we may view the domain of f as all of $\Delta(R)$, or as in a component G of $f^{-1}(F_0) \cap \Delta(R)$. Thus, the setting will determine the relevant collection of G_α's. Similarly, $n(r, \infty)$ will be the number of poles of f in either $\Delta(r)$ or $G(r)$. We will develop our method so that the reader can readily adapt it to either situation.

We make certain inessential normalizations: the β_k are pointwise disjoint, Γ_r meets each β_k at finitely many points, and none of the countably many branch points of f lies on any β_k. Finally, we assume that in each ball $B_\lambda(a_k, 3\eta)$ there is a line segment L_k passing through a_k such that relative to this ball, $\beta_{k-1} \cup \beta_k = L_k \backslash a_k$. When $k = q+1$, we take L to be the line constructed in Proposition 1.8. By making an arbitrarily small change in r, we may suppose that Γ_r does not pass through any of the a_k.

(2.1) Princple of the proof. The significance of length-area is seen from elementary considerations. The work that follows is to force the hypotheses of Lemma 2.2 to be satisfied.

(2.2) LEMMA. *Let the G_α be as above, and suppose G_α meets $B(r)$ in $P(\alpha)$ points $\zeta_{j,\alpha}$ whose images on S^2 are separated by some $\eta > 0$. Then*

$$L(r) \geq h \sum_\alpha (P(\alpha) - 2)^+. \tag{2.3}$$

Proof. Consider a fixed G_α, and $\zeta_{1,\alpha}, \ldots, \zeta_{P(\alpha),\alpha}$ on $\bar{G}_\alpha \cap B(r)$, such that $\lambda(\zeta_{i,\alpha}, \zeta_{j,\alpha}) > C\eta$; here the ζ's are listed in the order encountered on circuiting $B(r)$ in the positive direction. Since each G_α is connected, the ζ's are endpoints of $P(\alpha)$ disjoint arcs I of $B(r)$ and hence give a contribution at least $hP(\alpha)$ to $L(r)$. Then if $G_{\alpha'}$ is any other region determined by the $\{\beta_j\}$, $G_{\alpha'}$ must lie in one of the complementary domains of $\Delta(r)\backslash G_\alpha$.

Hence, given an initial choice of $G_{\alpha 1}$, choose $G_{\alpha 2}$ so that $G_{\alpha 2}$ is closest to G_{α_1} in one of these domains (there is not a unique such $G_{\alpha 2}$; in fact there are usually $P(\alpha)$ such). Then the closures of $G_{\alpha 2}$ and $G_{\alpha 1}$ can have at most two points in common on $B(r)$. Thus, $G_{\alpha 2}$ adds a term $P(\alpha_2) - 2$ to $L(r)$, since we are forced to introduce at least $P(\alpha_2) - 2$ new arcs I due to G_2. We exhaust the $\{G_\alpha\}$ in this manner, and (2.3) follows.

(2.4) The argument principle. Now for a fixed $k, 1 \leq k \leq q$, let $\beta(k)$ be the curve $\bigcup_k^q \beta_j$, so that $\beta(k)$ is a Jordan arc on S^2 which

joins a_k to ∞. Note that $\beta(1) \supset \beta(2) \supset \dots$. We also set $\beta'(k) = \beta_{q+1} \cup \{\bigcup_{1}^{k-1} \beta_j\}$. Choose a fixed θ, say $\theta = 0$, such that $f(re^{i\theta}) \notin \bigcup_j \beta_j$. Consider stopping times $\Theta(k) : 0 < \theta_1 < \theta_2 < \cdots < \theta_n < \theta_1 + 2\pi, n = n(k)$, such that $f(re^{i\theta_i}) \in \beta(k)$; we do not indicate the dependence on k of the θ's. This divides Γ_r into a union of arcs $\Gamma_i = \Gamma_i^k (1 \le i \le n)$ each of which starts and ends on $\beta(k)$; Γ_i^k is the image of $\theta_i \le t \le \theta_{i+1}$.

We partition the Γ_i^k into classes (I_k), (II_k) and (III_k):

(I_k) those arcs which lie completely in $B_\lambda(a_k, 2\eta)$,
(II_k) those arcs which lie completely in $B_\lambda(\infty, 2\eta)$,
(III_k) the others.

If γ is any curve (not necessarily closed) which does not pass through a_k or ∞, we set

$$\nu_k(\gamma) = \frac{1}{2\pi} \Delta_\gamma \arg(w - a_k), \tag{2.5}$$

and note that the normalization (1.10) reduces (1.5) to an estimate from below of $\sum_k \nu_k(\Gamma_r)$. If $G_\alpha = f^{-1}(F_\alpha)$, one of the subregions of $\Delta(r)$ determined by the $\{\beta_j\}$ as at beginning of this §, we let

$$\nu_k(\partial F_\alpha) = \frac{1}{2\pi} \sum_\gamma \Delta_\gamma \arg(w - a_k), \tag{2.6}$$

where the sum in (2.6) is over the f-images γ of the arcs of $\partial G_\alpha \cap B(r)$ (i.e., the *relative boundary* of G_α).

It is obvious that for curves Γ_i^k in classes (I_k) and (II_k) there can be no way to bound $v_k(\Gamma_i^k)$ in terms of the length $L(\Gamma_i^k)$. However we have

(2.7) Proposition. *Suppose f is such that $w^* = \infty$ satisfies the conditions of Proposition 1.8. Then for $1 \le k \le q$*

$$|\nu_k(\Gamma_r) - \sum_{(I_k)} \nu_k(\Gamma_i^k)| < hL(r). \tag{2.8}$$

Thus, the significant contributions to $\nu_k(\Gamma_r)$ arise from curves whose image winds about a_k and are close to a_k. We begin the proof here, and complete it in (2.13) below.

Proof. The critical case is when $i \in (III_k)$. Choose a (maximal) chain X of length p, $i_{\ell+1}, \ldots, i_{\ell+p}$ such that each $\Gamma^k_{\ell+j} \in (III_k)$. Let Γ be the portion of Γ_r, which corresponds to X; i.e., the image of $\theta_{\ell+1} < t < \theta_{\ell+p+1}$.

(2.9) LEMMA. *Let* Γ *be as above. Then*

$$|\nu_k(\Gamma)| \leq hL(\Gamma). \tag{2.10}$$

Proof of (2.10). Let $S(k)$ be S^2 with the open disks $\Delta_\lambda(a_k, \eta)$ and $\Delta_\lambda(\infty, \eta)$ deleted. Then for each k $S(k)$ is compact, so there exist $\sigma > 0$ and $M < \infty$ such that if γ is a continuum which meets $\beta(k), \beta'(k)$ and intersects $S(k)$, then

$$\lambda(\gamma) > \sigma \tag{2.11}$$

and, for any choice of argument on $S(k) \cap \beta(k)$,

$$|\sup \arg(w - a_k) - \inf \arg(w' - a_k)| < M \quad (w, w' \in S(k) \cap \beta(k)).$$

It is clear that by increasing M by at most 4π, we have a similar bound when w and w' are in $S(k) \cap \beta'(k)$.

Let Q be the number of i such that a subcurve of Γ^k_i meets $\beta'(k)$. Then it follows from the definition of M that

$$|\nu_k(\Gamma)| < 2Q + 2M :$$

we think of Γ having an initial and terminal portion which does not meet $\beta'(k)$, and then Q intermediate portions which join $\beta(k)$ to itself, passing through $\beta'(k)$. Similarly, $L(\Gamma) > Q\sigma$, so that (2.10) holds in the weaker form

$$|\nu_k(\Gamma)| \leq hL(\Gamma) + M. \tag{2.12}$$

It is possible to delete M in (2.12). If $L(\Gamma) > \eta$, it is obvious that M in (2.12) may be absorbed in the term $hL(\Gamma)$; if $L(\Gamma) < \eta$ and Γ meets $B_\lambda(a_k, \eta)$ or $B_\lambda(\infty, \eta)$, then Γ is a curve both of whose endpoints are on $\beta_k \cap B_\lambda(a_k, 3\eta)$ or $\beta_k \cap B_\lambda(\infty, 3\eta)$. In either case,

β_k is a ray eminating from a_k or ∞ in this region and so $v_k(\Gamma) \equiv 0$. Finally, if $L(\Gamma) < \eta$ but $\Gamma \cap \{B_\lambda(a_k, \eta) \cup B_\lambda(\infty, \eta)\} = \emptyset$, we see that in this case $\Gamma \equiv \Gamma_r$, a closed curve, so that $v_k(\Gamma) = 0$. Hence (2.10) holds in all cases.

(2.13) Completion of Proof of Proposition 2.7. By the normalization (1.10) with Proposition 1.8, it is clear that $\sum_{(II_k)} |\nu_k(\Gamma_i^k)| < KL(r)$. Thus (2.8) is a consequence of this and (2.10).

(2.14) An extension of Proposition 2.7. By (2.8), the significant contribution to $\nu_k(\Gamma_r)$ arises from portions of Γ_r which circuit a_k in a full revolution, and are contained in $B_\lambda(a_k, 2\eta)$. This can be made a bit sharper.

(2.15) LEMMA. *For each $k \in \{1, \ldots, q\}$, let Λ_i^k be the arcs of Γ_r which lie in $B_\lambda(a_k, 2\eta)$ and join β_{k-1} and β_k. Then*

$$|\nu_k(\Gamma_r) - \sum_i \nu_k(\Lambda_i^k)| < hL(r). \tag{2.16}$$

Proof. This follows at once from Proposition 2.7 and the observation that each arc Γ_i^k of that Proposition contains two arcs Λ_i^k (one which is mapped into F', one into F'') plus, perhaps, additional subarcs which start and end on one of β_{k-1} or β_k. Since the β's are radial segments in $B_\lambda(a_k, 3\eta)$, the latter arcs contribute nothing to $\nu_k(\Gamma_r)$ or $\nu_k(\Gamma_i^k)$.

(2.17) More on the role of ∞. We modify (2.5) and (2.6) to

$$\nu_k^*(\gamma) = \begin{cases} \nu_k(\gamma) & \text{if } \gamma \subset B_\lambda(a_k, 2\eta) \\ 0 & \text{otherwise} \end{cases} \tag{2.18}$$

and a similar interpretation for $\nu_k^*(\partial F_\alpha)$ (see (2.6)).

Let $w^* = \infty$ be normal in the sense of Proposition 1.8. We show that ∞ is typical in a very strong sense.

(2.19) LEMMA (I). *Let ∞ be normal. For each α, let $n(\alpha)$ be the number of poles on ∂G_α, so that ∂G_α is partitioned into $n(\alpha)$ components $\Gamma(\alpha, \beta)$. Then with the exception of a set of B poles with*

$$\sum_{(\alpha,\beta)\in B} 1 < hL(r), \tag{2.20}$$

the following is true. For each $k, 1 \leq k \leq q$, the number of disjoint arcs $\gamma \subset U_{\beta}\Gamma(\alpha, \beta)$ with

$$\nu_k^*(\gamma) = -\frac{1}{2} \tag{2.21}$$

plus the number of solutions to the equation

$$f(z) = a_k, \qquad z \in U_{\beta}\Gamma(\alpha, \beta) \tag{2.22}$$

equals $n(\alpha)$.

(II) Conversely, let $F_{\alpha} = f(G_{\alpha})$ be given, choose k as in (I), and suppose that $F_{\alpha} \not\subset B_{\lambda}(a_k, 3\eta)$. Let the $\{\Gamma(\alpha, \beta)\}$ be as in (I). Then with the exception of a set of B of (α, β) as in (2.20), each $\Gamma(\alpha, \beta)$ contains a pole.

(2.23) Remark. This lemma complements Miles [8], which covers the situation that $F_{\alpha} \subset B_{\lambda}(a_k, 3\eta)$ for many α; then there may be many a_k-values not compensated by poles In this situation, $\nu_k(\gamma) > 0$ for many arcs $\gamma \subset \Gamma_r$, in contradistinction to (2.21).

Proof. Choose k as above. If the f-image of a $\Gamma(\alpha, \beta)$ does not pass through a_k, then $\Gamma(\alpha, \beta)$ contains an arc of $B(r)$ whose image γ separates ∞ from a_k in F_{α}. Unless $\gamma \subset B_{\lambda}(a, 3\eta), (a = a_k, \infty)$, the argument of (2.11) shows that $\lambda(\gamma) > \sigma_1 > 0$, independent of α, k. By our normalization (1.10), the total number of poles so separated as α, β, k vary satisfies (2.20). This proves (I).

Conversely, let $n_1(\alpha, k), n_2(\alpha, k)$ be the number of solutions to (2.21) and (2.22) for a given α, and circuit ∂F_{α}. The arcs and a_k-values of (2.21) and (2.22) divide ∂F_{α} into $n(\alpha, k) = n_1(\alpha, k) + n_2(\alpha, k)$ portions $\Gamma(\alpha, \beta, k)$. To each $\Gamma(\alpha, \beta, k)$ which does not pass through $a_{q+1} = \infty$ corresponds a crosscut $\gamma = \gamma(\alpha, \beta, k)$ which separates ∞ from a_k. Since Proposition 1.8 holds, the argument of the paragraph immediately above shows that the number of such (α, β) can be absorbed in (2.20). This completes the proof.

(2.24) COROLLARY. *Let N^* be the number of pairs (α, β) which satisfy the hypotheses of Part (II) of Lemma 2.19. Then, if ∞ is normal in the sense of Proposition 1.8, we have*

$$|N^* - 2n(\infty)| < hL(r). \tag{2.25}$$

In particular, if P is the number of poles which are taken in these $\Gamma(\alpha, \beta)$, then

$$P \leq 2n(r, \infty) + hL(r). \tag{2.26}$$

Proof. The poles of f correspond to regions F_α which have ∞ in their closure. Hence Lemma 2.19 applies. The first part of the corollary now follows since each pole is on the boundary of two G_α's. Estimate (2.26) is immediate.

3. Proof of (1.5)

Let $n = n(\alpha)$ be the number of poles of f on G_α relative to $\Delta(r)$. Note that

$$\sum_\alpha \sum_\beta 1 = \sum_\alpha n(\alpha) = 2n(r, \infty), \tag{3.1}$$

and that the contribution of the exceptional (α, β) satisfies (2.20). Let the $\{\Gamma(\alpha, \beta)\}$ be as in Lemma 2.19. Choose $k \in \{1, \ldots, q\}$. If the f-image of a $\Gamma(\alpha, \beta)$ does not pass through a_k, then there is an arc $\gamma \subset \Gamma(\alpha, \beta) \cap B(r)$ whose f-image separates ∞ from a_k in F_α. Unless the image of γ lies in $B_\lambda(a_k, 3\eta) \cup B_\lambda(\infty, 3\eta)$, the argument of (2.11) shows that $\lambda(\gamma) > \sigma_1 > 0$, independent of α or k. We now apply Lemma 2.2 to each of these $n(\alpha)$ sets $\Gamma(\alpha, \beta)$. Let $P(\alpha, \beta)$ be the number of $k \in \{1, \ldots, q\}$ such that, as in (2.15), $\nu_k(\Lambda_i^k) < 0$ for an arc Λ_i^k of $\Gamma(\alpha, \beta)$, and $P(\alpha) = \Sigma_\beta P(\alpha, \beta)$. Then $\nu_k(\Lambda_i^k) = \nu_k^*(\Lambda_i^k) = -\frac{1}{2}$, so by (2.16), (3.1), (2.3) and (1.10) we have

$$\begin{aligned} \sum_k \nu_k(\Gamma_r) &\geq \sum_{k,\alpha} \nu_k(\Lambda_i^k) - hL(r) \\ &= -\frac{1}{2} \sum_\alpha \sum_\beta P(\alpha, \beta) - hL(r) \\ &= -\sum_\alpha n(\alpha) - \frac{1}{2} \sum_\alpha \sum_\beta \{P(\alpha, \beta) - 2\} - hL(r) \\ &\geq -\sum_\alpha n(\alpha) - \frac{1}{2} \sum_\alpha \sum_\beta \{P(\alpha, \beta) - 2)^+ - hL(r) \\ &= -2n(r, \infty) - hL(r) \\ &\geq -2S(r) - hL(r). \end{aligned} \tag{3.2}$$

By the argument principle, this is (1.5).

4. Proof of (1.6). Recall the discussion of Euler characteristic in, say, [5, pp. 135–7], [9, pp. 322–3]. We surround each $a_k (1 \leq k \leq q)$ by a small disk D_k, such that $\lambda(\zeta, a_k) \sim \eta$ for $\zeta \in \partial D_k$, and let $F_0 = S^2 \setminus \cup D_k$. Thus, there are now q crosscuts β_i; what is now β_q consists of what in § 2 had been a connected piece of β_q and β_{q+1} which passes through ∞. We estimate $\rho(\mathfrak{F})$ by the standard combinatorial inequality [5, p. 137], [9, p. 333]

$$\rho(\mathfrak{F}) \geq n - N \tag{4.1}$$

where N is the total number of domains $G_\alpha = f^{-1}(F_\alpha)$ and n is the number of crosscuts $\cup_j f^{-1}(\beta_j)$.

(4.2) Remark. Each crosscut γ bounds two domains $\{G_\alpha\}$; we will use the fact, needed for (4.1), that crosscuts γ which disconnect $\mathfrak{F}$ make no net change to either side of (4.1)

Consider the arcs $\Gamma(\alpha, \beta)$ of Lemma 2.19. In the context here, $\Gamma(\alpha, \beta) \subset \partial G_\alpha$ and we write $F_\alpha = f(G_\alpha)$ (so that $F_\alpha \subset F$, with $F = F'$ or F'', where F' and F'' are now bounded by the $\{\beta_k\}$ and portions of the $\cup_1^q \partial D_k$.

If $\Gamma \in \Gamma(\alpha, \beta)$, let $\Gamma^* = \Gamma \cap \Gamma_r$ be the portion of Γ in the relative boundary of $\mathfrak{F}$; this convention of starring will be used below. As in §2, choose $\eta > 0$ such that, if $j \neq k$, then $\lambda(\beta_j, \beta_k) > 100\eta$ (distance relative to $\mathfrak{F}$). The number of pairs (α, β) with $\lambda(\Gamma^*(\alpha, \beta)) > \eta$ is at most $hL(r)$. We place these exceptional pairs (α, β) into class *(I)*; by Lemma 1.8 and the normalization (1.10) we may also include in *(I)* all $\Gamma(\alpha, \beta)$ such that an endpoint of some $\Gamma^*(\alpha, \beta)$ lies in an η-neighborhood of ∞. Thus ($^\#\{E\}$ is the cardinality of E)

$$^\#\{(\alpha, \beta) \in (I)\} < hL(r). \tag{4.3}$$

We now introduce a significant set $\mathcal{G}$ of pairs $\{\alpha, \beta\}$, which are not in *(I)*. Let $\partial_0\mathfrak{F}$ be the *outer boundary* of $\mathfrak{F}$, i.e., the component of $\partial\mathfrak{F}$ which intersects Γ_r; thus $\partial_0\mathfrak{F}$ is connected and consists of part of Γ_r and perhaps arcs which are mapped to ∂D_k for various k. If $(\alpha, \beta) \notin (I)$, let (cf. Lemma 2.2) $p = p(\alpha, \beta)$ be a maximum choice of points $\{\zeta\}$ on $\partial_0\mathfrak{F} \cap \Gamma^*(\alpha,\beta)$ such that $\lambda(\zeta,\zeta') \geq 3\eta$. By Proposition 1.8 and the definition of *(I)*, we may assume that none of the ζ are in $B_\lambda(\infty, 3\eta)$. The class $\mathcal{G}$ will consist of all (α,β) for which $p \neq 1$ and a certain subset of the (α,β) for which $p = 1$.

If $p = 0$ then $\Gamma^*(\alpha,\beta) = \phi$ so that we have q crosscuts of $\mathfrak{F}$ corresponding to this (α,β), and none of these crosscuts disconnect $\mathfrak{F}$.

Now let $(\alpha,\beta) \notin (I)$ with $p = 1$, corresponding to a choice $\zeta = \zeta_1$. By hypothesis, $\Gamma^*(\alpha,\beta)$ contains a cross-cut γ^* of F_α which is contained in an η-neighborhood Ω of ζ_1. There are two possibilities. Since F_α is connected, it is easy to see that either $F_\alpha \subset \Omega$ or $F_\alpha \supset \{F \backslash \Omega\}$. Since $(\alpha,\beta) \notin (I)$, there is some k such that the two endpoints of each component γ^* of $\Gamma(\alpha,\beta)$ are contained in β_k, D_k or D_{k+1}. Hence, when $F_\alpha \subset \Omega$, we see from (4.2) that *all* crosscuts 'over' $\cup\beta_k$ can be ignored in computing (4.1), since each disconnects $\mathfrak{F}$. This is important since there can be no upper bound for the number of such components $\Gamma(\alpha,\beta)$. The remaining pairs (α,β) for which $p = 1$ are assigned to $\mathcal{G}$. In this case, $F_\alpha \supset F\backslash\Omega$, and since $p = 1$, it follows that $\Gamma(\alpha,\beta)$ will contain q crosscuts of $\mathfrak{F}$ which terminate at each D_k, and so there are at least $q - 1$ crosscuts which do not separate $\mathfrak{F}$, since at least $q - 1$ cannot meet $\Gamma^*(\alpha,\beta)$.

Finally, if $p(\alpha,\beta) = p \geq 2$ and $(\alpha,\beta) \notin (I)$, we see that $F_\alpha \supset F \backslash \cup_1^p \{B_\lambda(\zeta_i, 3\eta)\}$. In this situation, there are again q crosscuts from $\Gamma(\alpha,\beta)$, but we are assured only that $q - p$ do not disconnect $\mathfrak{F}$. However, by Lemma 2.2,

$$\sum_{\{(\alpha,\beta)\}\subset\mathcal{G}} \{p(\alpha,\beta) - 2\}^+ < hL(r). \tag{4.4}$$

If $\mathcal{G}$ is as defined above, it follows from Proposition (1.8) and (1.10) that (3.1) holds. Let $n_{\mathcal{G}}$ and $N_{\mathcal{G}}$ be the contribution to n and N in (4.1) which arise from $\{(\alpha,\beta) \in \mathcal{G}\}$. Since ∞ lies on each $\Gamma(\alpha,\beta)$ if $(\alpha,\beta) \in \mathcal{G}$, we deduce from our definition of (*I*), (4.3), (3.1) and (1.8) that

$$
\begin{aligned}
n - N &\geq n_{\mathcal{G}} - N_{\mathcal{G}} - hL(r) \\
&\geq \frac{1}{2}[q^{\#}\{\mathcal{G} \cap \{p = 0\}\} + (q-1)^{\#}\{\mathcal{G} \cap \{p = 1\}\} \\
&\quad + \sum_{m \geq 2} (q - m)^{\#}\{\mathcal{G} \cap \{p = m\}\}] - hL(r) \\
&\geq \sum_{q \leq 2} (q - p) + \sum_{p \geq 2} (q - 2) - \sum_{\mathcal{G}} (p - 2)^{+} - hL(r) \\
&\geq \frac{1}{2}(q - 2)^{\#}\mathcal{G} - \sum_{\mathcal{G}} (p - 2)^{+} - hL(r) \\
&\geq (q - 2)n(r, \infty) - hL(r) \\
&= (q - 2)S(r) - hL(r).
\end{aligned}
$$

Since $\rho_0 = q - 2$, we have proved (1.6).

References

[1] Ahlfors, L. V., *Zur Theorie der Überlagerungsflächen*, Acta Math., 65 (1935), pp. 157–194.

[2] ———, *Über die Anwendung Differentialgeometrischer Methoden zur Untersuchung von Überlagerungsflächen*, Acta Soc. Sci. Fenn. New Series A, V. II(6), pp. 1–17.

[3] ———, *Collected Papers, Vol. 1*, Birkhäuser, Boston, 1982.

[4] Ahlfors, L. V., and Sario, *Riemann Surfaces*, Princeton, Princeton, 1960.

[5] Hayman, W. K., *Meromorphic Functions*, Oxford, 1964.

[6] Lyzzak, A. K. and Stephenson, K., *The structure of open continuous functions having two valences*, Trans. Amer. Math. Soc. 327 (1991) pp. 525–566.

[7] Miles, J., *A note on Ahlfors' theory of covering surfaces*, Proc. Amer. Math. Soc., 21 (1969), pp. 30–32.

[8] ———, *Bounds on the ratio $n(r, a)/S(r)$ for meromorphic functions*, Trans. Amer. Math. Soc., 162 (1971), pp. 383–393.

[9] Nevanlinna, R., *Analytic Functions*, Springer-Verlag, New York, 1970.

[10] Pesonen, M., *A path family approach to Ahlfors' value-distribution theory*, Ann. Acad. Sci, Fenn, Ser. A, I Math. Dissertationes 39 (1982), pp. 1–32.

[11] S. Stoïlow, *Principes Topologiques de la Théorie des Fonctions Analytiques*, Gauthier-Villars, Paris, 1938.

[12] Y. Toki, *Proof of Ahlfors principal covering theorem*, Rev. Math. Pures et Appl., 2 (1957), pp. 277–280.

$\mathbb{C}^n$-CAPACITY AND MULTIDIMENSIONAL MOMENT PROBLEM

G. M. Henkin, A. A. Shananin
(Moscow)

Introduction

Let K be a compact set in the n-dimensional complex space $\mathbb{C}^n$, $H(K)$ be a space of holomorphic functions on K, $H'(K)$ be the space of linear continuous functionals over $H(K)$. We will write down the value of the functional $\mu \in H'(K)$ on the function $h \in H(K)$ in the form of $<\mu, h>$. The numbers of the form $C_\nu(\mu) = <\mu, Z^\nu>$ are called the moments of the analytical functional μ, where $Z^\nu = Z_1^{\nu_1} \dots Z_n^{\nu_n}$ is a holomorphic monomial of the degree $|\nu| = \nu_1 + \dots + \nu_n$; $Z = (Z_1, \dots, Z_n) \in \mathbb{C}^n$, $\nu = (\nu_1, \dots, \nu_n) \in \mathbb{Z}_+^n$.

The problem arising from a number of applications (computational tomography [1], inverse problem of the potential theory [2], quadrature formulae [3], and even production functions theory [4]) is to reconstruct a functional from $H'(K)$ through its moments.

The necessary and sufficient condition of uniqueness of a functional $\mu \in H'(K)$, which has the fixed moments $\{C_\nu(\mu)\}$ is polynomial convexity of the compact set K, since polynomial convexity of K is necessary and sufficient in order that any function from $H(K)$ will be approximated by holomorphic polynomials (A. Weil, 1932).

If a functional μ is given by positive measure on the compact set $K \subset R^n \subset \mathbb{C}^n$ then the considered problem is called the classical moment problem. This classical problem is effectively and completely solved only for the case $n = 1$ (see [5]).

In connection with applications the problem of the approximate reconstruction of the functional $\mu \in H'(K)$ through the finite number of moments C_ν, $|\nu| \leq N$ is of particular interest. In the classical theory this problem is called the Markov moment problem. In order to solve this problem it is necessary to answer at least the following questions:

1. What is a guaranteed estimate of the accuracy of the possible reconstruction of the functional $\mu \in H'(K)$ if the moments $C_\nu(\mu)$, $|\nu| \leq N$ and certain norm of the functional μ are known?

2. How to find actually the functional $\mu \in H'(K)$ with a priori given moment $C_\nu(\mu), |\nu| \leq N$ and with some suitable norm?

It turned out that these questions are closely connected with several modern themes from several complex variables.

Namely, for exact answer to the question 1, it is used the results of the theory of extremal plurisubharmonic functions and of the complex Monge-Ampère equation on the parabolic manifolds obtained in the papers [6]–[22] and also the theory of the Fantappie-Martineau analytical functional [23]–[27]. The modern variants of the interpolational formulae of the Jacobi type for the holomorphic functions in the hyperconvex domains [28], [29] are very useful for the answer to the question 2.

In this article we give a suitable answer to the question 1 and indicate the simplest applications. The constructive answer to the question 2 will be given in the other paper.

§1. The results.

The compact subset $K \subset \mathbb{C}^n$ is called regular (see [8], [9], [14]) if there exists (and unique) a continuous solution U_K of the following exterior Dirichlet problem for the complex Monge-Ampère equation: $U_K(Z)$ is a plurisubharmonic function in $\mathbb{C}^n \backslash K$,

$$\begin{aligned} &\det\left[\frac{\partial^2 U_K}{\partial Z_\alpha \partial \bar{Z}_\beta}(Z)\right] = 0 && \text{in } \mathbb{C}^n \backslash K \\ &U_K(Z) = \log|Z| + O(1) && \text{as } |Z| \to \infty \\ &U_K(Z) = 0 && \text{if } Z \in \partial K. \end{aligned} \tag{1.1}$$

The compact subset $K \subset \mathbb{C}^n$ is called (see [23]–[26]) linear convex if for any point $W \in \mathbb{C}^n \backslash K$ a set of complex hyperplanes passing through W and not crossing K is non-empty and contractible.

The compact K is called strictly linear convex if its boundary ∂K is smooth and for any point $W \in \partial K$ the complex tangent hyperplane $T_W^C(\partial K)$ have the unique point of contact $\{W\}$ with ∂K and this contact not higher than the first order. Any linear convex compact set K may be represented in the form of

$$K = \bigcap_{j=1}^{\infty} K_j,$$

where $K_1 \supset K_2 \supset \ldots$ is a sequence of strictly linear convex compact sets. Besides, there takes place the monotonic convergence for regular linear convex compact sets K

$$U_j(Z) \to U_K(Z) \qquad \text{for } j \to \infty, Z \in \mathbb{C}^n \backslash K, \tag{1.2}$$

where $U_j(Z) = U_{K_j}(Z)$- smooth solutions of the type (1.1) of the Monge-Ampère equation in $\mathbb{C}^n \backslash K_j$. Existence and uniqueness of such solutions for strictly linear convex compact sets is proved in [17].

We suppose without loss of generality that a linear convex compact K contains the origin of coordinates in $\mathbb{C}^n$. We define a domainK' dual to the compact set K by the formula

$$K' = \{p \in (\mathbb{C}^n)' : pZ + 1 \neq 0 \qquad \text{for } Z \in K\}.$$

For the domain K' we have such a representation

$$K' = \bigcup_{j=1}^{\infty} K'_j,$$

where $K'_1 \subset K'_2 \subset \ldots$ is a sequence of strictly linear convex domains dual to K'_j.

According to Lempert [11], [12] there exist smooth solutions $V_j = V_{K'_j}$ of the Monge-Ampère equations in the domains $K'_j \backslash \{0\}$: $V_j(p)$ is a plurisubharmonic function in $K'_j \backslash \{0\}$

$$\begin{array}{ll} \det \left[\frac{\partial^2 V_j(p)}{\partial \bar{p}_\alpha \partial p_\beta}\right] = 0 & \text{in } K'_j | \{0\} \\ V_j(p) = \log |p| + O(1) & \text{for } p \to 0 \\ V_j(p) = 0 & \text{for } p \in \partial K'_j. \end{array} \tag{1.3}$$

Besides, $O(1) = S'_j(\frac{p}{|p|}) + O_j(|p|)$, where S'_j is a smooth function on $\mathbb{C}P^{n-1}$, i.e., $S'_j(\lambda \cdot p) = S'_j(p), \forall \lambda \in \mathbb{C}$.

The following nice formula is valid ([17], p. 882)

$$V_j(p) = -U_j(Z(p)), \tag{1.4}$$

where

$$Z(p) = \frac{\partial V_j(p)}{\partial p}\left(p\frac{\partial V_j(p)}{\partial p}\right)^{-1}$$

is a diffeomorphism of the domain $K_j'\backslash\{0\}$ on $\mathbb{C}^n\backslash K_j$;

$$\frac{\partial V_j}{\partial p} = \left(\frac{\partial V_j}{\partial p_1}, \dots, \frac{\partial V_j}{\partial p_n}\right).$$

It follows from (1.2), (1.4), in particular, that there takes place a monotonic convergence

$$V_j(p) \to V_{K'}(p), \quad j \to \infty, \quad p \in K'\backslash\{0\}, \tag{1.5}$$

where $V_{K'}(p)$ is a continuous solution of the Monge-Ampère equation of the type (1.3) in the domain $K'\backslash\{0\}$.

For regular linear convex compact sets K so called [16], [22] Robin functions of the compact set K and of the domain K' are defined and continuous on $\mathbb{C}P^{n-1}$

$$\begin{aligned} S(\zeta) &= \limsup_{\lambda\to\infty}(U_K(\lambda\zeta) - \log|\lambda|) \\ S'(\zeta) &= \limsup_{\lambda\to 0}(V_{K'}(\lambda\zeta) - \log|\lambda|), \end{aligned} \tag{1.6}$$

where $\zeta \in \mathbb{C}^n : |\zeta| = 1$ is identified with a point of $\mathbb{C}P^{n-1}$.

Following Lelong [21] we shall call the functions $\gamma(\zeta) = \exp(-S(\zeta))$ and $\gamma'(\zeta) = \exp(-S'(\zeta))$, $\zeta \in \mathbb{C}P^{n-1}$ capacitative indicatrices of the compact K and of the domain K' respectively.

Due to the statement of convergence of the Robin functions from Bedford-Taylor ([22], p. 163) it follows from (1.2) and (1.5) that

$$\begin{aligned} \gamma_j(\zeta) &\to \gamma(\zeta), \quad j \to \infty, \quad \zeta \in \mathbb{C}P^{n-1} \\ \gamma_j'(\zeta) &\to \gamma'(\zeta), \quad j \to \infty, \quad \zeta \in \mathbb{C}P^{n-1}, \end{aligned} \tag{1.7}$$

where γ_j and γ_j' are capacitative indicatrices of the compact set K_j and of the domain K_j' respectively.

The following explicit relation between indicatrices γ and γ' implies from (1.3), (1.4), (1.6), (1.7)

$$\gamma(\bar{\zeta} - \frac{\partial}{\partial \zeta} \ln(\gamma'(\zeta))^2) = \frac{|\bar{\zeta} - \frac{\partial}{\partial \zeta} \ln(\gamma'(\zeta))^2|}{\gamma'(\zeta)}, \tag{1.8}$$

where $\zeta \in \mathbb{C}^n : |\zeta| = 1$.

The most important examples of linear convex and simultaneously regular compact sets are compact sets in $\mathbb{C}^n$, which are closures of the bounded linear-convex domains in $\mathbb{C}^n$ with smooth boundary or closures of the bounded convex domains in $R^n \subset \mathbb{C}^n$. In particular, for the complex ball $K = \{Z \in \mathbb{C}^n : |Z| \leq R\}$ it is well known that $U_K(Z) = \ln \frac{|Z|}{R}$. W. Stoll [10] obtained necessary and sufficient property of $U_K(Z)$ which characterizes the manifolds equivalent to the complex ball. For the real ball $K = \{Z = x + iy \in \mathbb{C}^n : |x| \leq R, y = 0\}$ M. Lundin [19] obtained the following nice formula $sh^2 U_K(Z) = \frac{1}{2}(|Z|^2 - R^2 + |Z^2 - R^2|)$.

The entire function $\hat{\mu}(\zeta)$ of the variable $\zeta \in \mathbb{C}^n$ of the form

$$\hat{\mu}(\zeta) = <\mu, \exp(i\zeta \cdot Z)>, \tag{1.9}$$

where $\zeta Z = \zeta_1 Z_1 + \ldots + \zeta_n Z_n$, is called the Fourier-Laplace transform of the analytical functional $\mu \in H'(K)$.

For the functional $\mu \in H'(K)$ where K is a regular compact set, we define semi-norms of the form

$$\|\mu\|_\delta = \sup_{h \in H(K_\delta) : \quad |h(Z)| \leq 1, \quad Z \in K_\delta,} |<\mu, h>| \tag{1.10}$$

where $K_\delta = \{Z \in \mathbb{C}^n : U_K(Z) \leq \delta\}$, $\delta > 0$, U_K satisfies (1.1).

The following result gives a sufficiently exact answer to the question 1 for functionals with support on the regular linear convex compact.

Theorem. *Let K be a regular linear convex compact in $\mathbb{C}^n$ and $\gamma'(\zeta)$ be a capacitative indicatrix of the domain K'. Then*

A) *for any $N \in \mathbb{Z}_+$ any functional $\mu \in H'(K)$ with the moments*

$C_\nu(\mu) = 0$ *for* $|\nu| \leq N$, *any* $\zeta \in \mathbb{C}^n, |\zeta| = 1$, *any* $\lambda \in \mathbb{C}$ *and for any* $\delta > 0$ *there takes place the following inequality*

$$|\hat{\mu}(\lambda\zeta)| \leq \frac{\|\mu\|\delta}{\mathrm{d}(K, K_\delta)} \left[\frac{e^{1+\delta}|\lambda| \left(1 + O_{K,\zeta}\left(\frac{|\lambda|}{N}\right)\right)}{\gamma'(\zeta) \cdot (N+1)} \right]^{N+1}, \quad (1.11)$$

where $O_{K,\zeta}(\varepsilon) \to 0$ if $\varepsilon \to 0$; $d(K_1 K_\delta) = \inf |1+p \cdot z|, z \in K, p \in K'$

B) *for any* $\zeta \in \mathbb{C}^n, |\zeta| = 1$, *any* $N \in \mathbb{Z}_+$ *there exists the functional* $\mu = \mu_{N,\zeta} \in H'(K)$ *with the moments* $C_\nu(\mu) = 0$ *for* $|\nu| \leq N$ *and with estimates of the form*

$$|\hat{\mu}(\lambda\zeta) \geq \frac{(n-1)!}{(N+n)!} \left(\frac{|\lambda|}{|\gamma'(\zeta)|}\right)^{N+1}$$

$$e^{-d_k(\zeta)\cdot|\lambda|} \left(1 - O_{K,\zeta}\left(\frac{|\lambda|^2}{\sqrt{N}}\right)\right) \quad (1.12)$$

$$\|\mu\|_\delta \leq \frac{(n-1)!}{(2\pi)^n} \int\limits_{Z \in \partial K_\delta} |\omega'(\eta(Z)) \wedge \omega(Z)|, \quad (1.13)$$

where $\eta(Z)$ *is any smooth* $\mathbb{C}^n$*-valued function of the variable* $Z \in \partial K_\delta$ *with the property* [26]: *for all* $Z \in \partial K_\delta$ *and* $W \in K$ *we have* $1 + \eta(Z) \cdot Z = 0$ *and* $1 + \eta(Z) \cdot W \neq 0$; $\omega(Z) = \bigwedge\limits_{j=1}^{n} dZ_j$; $\omega'(\eta) = \sum_{j=1}^{n} (-1)^j \eta_j \bigwedge\limits_{V \neq j} d\eta_j d_k(\zeta) > 0$.

For the case when the compact set K is a strictly linear convex then the compact set K_δ for any $\delta > 0$ is also strictly linear convex [17]. Using in this case $\delta = 0$ and $\eta(Z) = \frac{\partial U_k(Z)}{\partial Z} \bigg/ \left(Z \cdot \frac{\partial U_K(Z)}{\partial Z}\right)$ we obtain from (1.13) that the functionals $\mu_{N,\zeta}$ have a uniformly bounded norm $\|\mu\|_0$.

The theorem, roughly speaking, means that if the moments $C_\nu(\mu), |\nu| \leq N$ are known for the finite measure μ with the support on the K then its Fourier-Laplace transform $\hat{\mu}(\zeta)$ is reconstructed with accuracy of the order $\|\mu\| \left(\frac{e \cdot |\zeta|}{\gamma'\left(\frac{\zeta}{|\zeta|}\right)(N+1)}\right)^{N+1}$ and not better, in general. It is important to express capacitative indicatrix $\gamma'(\zeta/|\zeta|)$ in geometric terms in order to use such an estimate. For general case it is not simple. However, the following statement is valid for the particular case when the compact set K and direction ζ are real.

Proposition. *Let K be a closure of the bounded convex domain in $R^n \subset \mathbb{C}^n$. Then the following equality is valid*

$$(\gamma'(\zeta))^{-1} = \frac{1}{4}\left[\sup_{x\in K}(\zeta \cdot x) - \inf_{x\in K}(\zeta \cdot x)\right] \tag{1.14}$$

for any real $\zeta \in R^n, |\zeta| = 1$.

Remark. If we drop demand of the regularity of the linear convex compact set in the theorem then the theorem is still valid if we will write in the statement that $\zeta \in \mathbb{C}P^{n-1} \backslash E$ where E is some polar subset of $\mathbb{C}P^{n-1}$. In addition, instead of the function $U_K(Z)$ of the form (1.1) it is necessary to use extreme plurisubharmonic function [8], [9] of the form

$$U_K(Z) = \sup\{U(Z) : U \text{ is plurisubharmonic on } \mathbb{C}^n \backslash K\}$$
$$U(Z) \leq \log|Z| + O(1), U(Z) \leq 0 \text{ on } \partial K.$$

The necessary properties of the Robin function for such extremal functions are obtained by P. Lelong [21] and E. Bedford, B. Taylor [22].

This theorem supposes may be more clear interpretation in terms of the best approximations of the function $\exp(\zeta \cdot Z)$ by polynomials on the compact set K.

Let us define the numbers

$$E_N(K, f) = \inf_{P_N} \sup_{Z\in K} |f(Z) - P_N(Z)|,$$

where P_N is a polynomial of the degree N in $Z = (Z_1, \dots, Z_n)$.

Consequence 1. *The following equality takes place*

$$\lim_{N\to\infty} N \cdot E_N^{1/N}(K, e^{i\zeta Z}) = e \cdot |\zeta| \Big/ \gamma'(\zeta/|\zeta|)$$

for any regular linear convex compact set $K \subset \mathbb{C}^n$ and any $\zeta \in \mathbb{C}^n$.

Note, that the result of the consequence 1 may be considered as complement of the following general approximating result of Siciak [6], [9]. In order that $f \in H(K_\delta)$ (see (1.10) it is necessary and sufficient that

$$\overline{\lim_{N\to\infty}} E_N^{1/N}(K, f) \le e^{-\delta}.$$

Now we will give an application of the theorem to one of computational tomography problem—to an estimate of the accuracy of the Radon transform inversion through the finite number of directions.

The transform of the type

$$R_\mu(\omega, s) = \frac{\partial}{\partial s} \int\limits_{\{x\in R^n : \omega x \le s\}} \mu(dx),$$

where $s \in R, \omega \in S^{n-1} = \{\omega \in R^n : |\omega| = 1\}$ is called the Radon transfrom for a finite measure μ with compact support in R^n.

The finite subset Ω of the sphere S^{n-1} is called N-solvable [1] if any polynomial $P_N(x)$ of the degree N is represented in the form

$$P_N(x) = \sum_{\omega\in\Omega} P_{N,\omega}(\omega \cdot x), \tag{1.15}$$

where $P_{N,\omega}$ is a polynomial of degree N of the variable $\omega \cdot x$. For the number of elements $|\Omega|$ in Ω we have the estimate

$$|\Omega| \ge C_{N+n-1}^{n-1}. \tag{1.16}$$

Conversely, if the inequality (1.16) is held and elements in Ω are in the general position then Ω is N-solvable (see [1]).

If the Radon transform $R_\mu(\omega, s), \omega \in \Omega$ is known for the measure μ and Ω is N-solvable, then the moments $C_\nu(\mu)$ of the order $|\nu| \le N$ are known for the measure μ due to (1.15).

Hence from the theorem we obtain the following consequence.

Consequence 2. *Let a support of the finite measure μ belong to the closure of the bounded convex domain $K \subset R^n$ and let the Radon transform $R_\mu(\omega, s)$ of the measure μ is equal to zero for directions ω belonging to N-solvable subset Ω. Then the Fourier-Laplace transform $\hat{\mu}(\zeta)$ for any $\zeta \in \mathbb{C}^n$ admits the estimate of the form (1.11).*

Note, that due to (1.14) we have $\gamma'(\zeta) = 2$ for the real unit sphere $K^1 = \{x \in R^n : |x| \leq 1\}$ and for real directions $\zeta \in S^{n-1}$. So, for this case the consequence 2 yields a preciser estimate:

$$\sup_{\{\zeta \in R^n : |\zeta| \leq \theta N\}} |\hat{\mu}(\zeta)| \leq \frac{O(1)}{\sqrt{N+1}} \cdot \left(\frac{\theta e}{2}\right)^{N+1} \cdot \|\mu\|_0$$

for any $\theta < 2/e$.

It is interesting to associate this result with the following Logan-Louis estimate (see [1]):
under the conditions of the consequence 2 we have

$$\int\limits_{\{\zeta \in iR^n : |\zeta| \leq \theta N\}} |\hat{\mu}(\zeta)| d\zeta \leq \beta(\theta) e^{-C(\theta)(N+1)} \|\mu\|_0$$

for $K = K^1$ and for any $\theta < 1$.

§2. The proof of the theorem.

This proof essentially uses the notion of the Fantappie indicatrix of the analytical functional.

The holomorphic function of the type

$$\Phi_\mu(p) = <\mu, \frac{1}{(1+pZ)}> \tag{2.1}$$

in the domain K' is called the Fantappie indicatrix of the analytical functional $\mu \in H'(K)$. Immediately from the definition (2.1) it follows that the equality $C_\nu(\mu) = <\mu, Z^\nu> = 0$ for $|\nu| \leq N$ is equivalent to the equalities

$$\Phi_\mu^{(\nu)}(0) = \frac{d^\nu \Phi_\mu}{dp_1^{\nu_1}, \dots, dp_n^{\nu_n}}(0) = 0, \tag{2.2}$$

for $|\nu| \leq N, \quad \nu = (\nu_1, \dots, \nu_n)$.

The Fantappie transform $\Phi_\mu(p)$ is simply expressed through the Fourier-Laplace transform $\hat{\mu}(\zeta)$

$$\Phi_\mu(p) = i \int_0^\infty e^{-i\tau} \hat{\mu}(\tau p) d\tau, \quad p \in K'. \tag{2.3}$$

Martineau [24] obtained a general formula expressing $\hat{\mu}(\zeta)$ through $\Phi_\mu(p)$ on the basis of the Cauchy-Fantappie-Leray formula (see [27], [29]). Here we will have a need of the following elementary formula.

$$\hat{\mu}(\zeta) = \frac{1}{2\pi i} \int_{\{\lambda \in \mathbb{C}: |\lambda| = R\}} \frac{1}{\lambda} e^{-i\lambda} \Phi_\mu \left(\frac{\zeta}{\lambda} \right) d\lambda, \tag{2.4}$$

where R is such that $\zeta/\lambda \in K'$ for any $\lambda : |\lambda| = R$.

The formula (2.4) is a simple consequence of the classical Cauchy formula. In fact, substituting the Cauchy representation

$$e^{i\zeta \cdot Z} = \frac{1}{2\pi i} \int_{\lambda \in \mathbb{C}: |\lambda| = R} \frac{e^\lambda d\lambda}{\lambda - i\zeta Z}$$

in the equality (1.9) we obtain

$$\hat{\mu}(\zeta) = \frac{1}{2\pi i} \int_{|\lambda| = R} e^\lambda \frac{d\lambda}{\lambda} < \mu, \frac{1}{1 - \frac{i\zeta Z}{\lambda}} > = \frac{1}{2\pi i} \int_{|\lambda| = R} \frac{e^\lambda d\lambda}{\lambda} \Phi_\mu \left(-\frac{i\zeta}{\lambda} \right).$$

The formula (2.4) allows to obtain necessary estimate for $\hat{\mu}(\zeta)$ on the basis of suitable estimates for $\Phi_\mu(\zeta/\lambda)$. We will obtain estimates for $\Phi_\mu(\zeta/\lambda)$ from equalities (2.2) and from the following immediate estimate.

$$\|\Phi_\mu(p)\| \leq \|\mu\|_\alpha \sup_{Z \in K_\alpha} \left| \frac{1}{1 + pZ} \right|, \tag{2.5}$$

where $\|\mu\|_\alpha$ is a norm of the form (1.10), $K_\alpha = \{Z \in \mathbb{C}^n : U_K(Z) \leq \alpha\}, \quad p \in K'$.

Suppose, further, $K_\delta = \{Z \in \mathbb{C}^n : U_K(Z) \leq \delta\}$, $K'_\delta = \{p \in K' : V_{K'}(p) + \delta < 0\}$, $\delta > 0$, where U_K and $V_{K'}$ are the functions satisfying (1.1), (1.5).

Consider now the plurisubharmonic function

$$\Psi_\delta(p) = \frac{1}{N+1} \ln \frac{|\Phi_\mu(p)| d(K, K_\delta)}{\|\mu\|_\delta}. \tag{2.6}$$

This function is negative in the domain $K'_\delta \subset K'$ due to (2.5). The estimate $\Psi_\delta(p) \le \ln|p| + O(1), \quad p \in K'_\delta$ also takes place due to (2.2). Due to (1.5) the function $V_{K'}(p) + \delta$ satisfies the Monge-Ampère equation (1.3) in the domain K'_δ. As it was shown in [18], [20] such a function is extremal plurisubharmonic function in the following sense:

$$\begin{aligned} V_{K'}(p) + \delta = \sup\{V/(p) : V \text{ is plurisubharmonic} \\ V(p) \le 0 \text{ and } V(p) \le \ln|p| + O(1) \text{ in } K'_\delta\} \end{aligned} \tag{2.7}$$

So, $\Psi_\delta(p) \le V_{K'}(p) + \delta$. From (2.6), (2.7) it follows that

$$|\Phi_\mu(p)| \le \frac{\|\mu\|_\delta}{d(K, K_\delta)} \exp\left[(N+1)(V_{K'}(p) + \delta)\right] \tag{2.8}$$

for $p \in K'_\delta$.

Substitute now the estimate (2.8) in the formula (2.4). Taking into account (1.3), (1.5) we obtain the following inequality

$$\begin{aligned} |\hat{\mu}(\zeta)| &\le e^R \frac{\|\mu\|_\delta}{d(K, K_\delta)} \exp\left[(N+1)(\sup_{\varphi \in [O, 2\pi]} V_{K'}\left(\frac{\zeta e^{i\varphi}}{R}\right) + \delta)\right] \\ &= e^R \frac{\|\mu\|_\delta}{d(K, K_\delta)} \exp\left[(N+1)(\log\left|\frac{\zeta}{R}\right| - \log\gamma'\left(\frac{\zeta}{|\zeta|}\right) + \delta \right. \\ &\quad \left. + O_{K, \frac{\zeta}{|\zeta|}}\left(\left|\frac{\zeta}{R}\right|\right)\right] \end{aligned}$$

for any $\zeta \in \mathbb{C}^n$, $\delta > 0$ and for such R that $\zeta R^{-1} e^{i\varphi} \in K'_\delta$ for all $\varphi \in [O, 2\pi]$. Suppose $R = N + 1$, we obtain

$$|\hat{\mu}(\zeta)| \le \frac{\|\mu\|_\delta}{d(K, K_\delta)} \left[\frac{e^{1+\delta}|\zeta|(1 + O_{K, \zeta/|\zeta|}\left(\left|\frac{\zeta}{N+1}\right|\right)}{\gamma'\left(\frac{\zeta}{|\zeta|}\right)(N+1)}\right]^{(N+1)}.$$

The estimate (1.11), i.e. the part A) of the theorem is proved.

In order to prove the part B) of the theorem we shall have need of one more formula for the capacitative indicatrix:

$$(\gamma'(\zeta))^{-1} = \sup_{\{F\in H(K'):F(0)=0,|F(\zeta)|\le 1,\zeta\in K'\}} \left|\zeta\frac{\partial F}{\partial \zeta}(0)\right|. \tag{2.9}$$

The proof of (2.9) is based on the Lempert results [15]. Due to (1.3), (1.5) for the solution $V(p)$ of the Monge-Ampère equation in the domain we have an asymptotic equality

$$V(\lambda\zeta) = \log|\lambda| - \log\gamma'(\zeta) + O(|\lambda|) \text{ for } \lambda \to 0, \tag{2.10}$$

where $p = \lambda\zeta, \zeta \in \mathbb{C}^n, |\zeta| = 1; \quad \lambda \in \mathbb{C}$.

Further, the following equality is valid (Lempert [15])

$$V_j(p) = \sup_{\{F\in H(K_j'):F(0)=0, \quad |F|\le 1\}} \ln|F(p)|, \tag{2.11}$$

where functions V_j satisfies (1.3).

Taking into account (1.5) from (2.11) we obtain also the equality

$$V(p) = \sup \ln|F(p)|. \tag{2.12}$$
$$\{F \in H(K') : F(0) = O, |F| \le 1\}$$

The equality (2.9) follows from (2.12).

Now we prove part B) of the theorem. We fix $\zeta \in \mathbb{C}^n : |\zeta| = 1$ and $N \in \mathbb{Z}_+$. Due to (2.10) there exists a function $F \in H(K')$ with the property

$$|F(p)| \le 1, \quad p \in K' \text{ and } F(\lambda\zeta) = (\gamma'(\zeta))^{-1}\lambda + O_{K,\zeta}(\lambda^2) \text{ for } \lambda \to 0 \tag{2.13}$$

Consider, further, a holomorphic function $\Psi(p) = F^{N+1}(p)$. We have

$$|\Psi(p)| \le 1, \quad p \in K' \text{ and}$$
$$\frac{\partial^\nu \Psi}{\partial p_1^{\nu_1}, \ldots, \partial p_n^{\nu_n}}(0) = 0 \text{ for } |\nu| \le N. \tag{2.14}$$

Due to the Martineau theorem [23], [24] refined in [25], [26], there exists a functional $\mu \in H'(K)$ such that its indicatrix $\Phi_\mu(p)$ satisfies the equality

$$\frac{1}{(n-1)!} D^{n-1} \Phi_\mu(p) = \Psi(p), \tag{2.15}$$

where $D\Phi = \Phi + p\frac{\partial \phi}{\partial p}$. It follows from (2.14), (2.15) that $C_\nu(\mu) = 0$ for $|\nu| \leq N$. We will prove the estimate (1.13).

Let,

$$L\mu = \frac{(n-1)!}{(2\pi i)^n} \Psi(\eta(Z)) \wedge \omega'(\eta(Z)) \wedge \omega(Z),$$

where $Z \to \eta(Z)$ is any smooth mapping with the property: for any $Z \in \partial K_\delta$ and $W \in K$ we have $1 + \eta(Z) \cdot Z = 0$ and $1 + \eta(Z)W \neq 0$. Let h be any bounded holomorphic function on K_δ. Due to the Cauchy-Fantappie-Leray formula we have (see [24]–[26]):

$$<\mu, h> = \int\limits_{Z \in \partial K_\delta} L\mu \wedge h. \tag{2.16}$$

The estimate (1.13) is an immediate consequence of (2.16). We will prove now the estimate (1.12). Taking into account formulae (2.4), (2.15) we obtain the equality

$$\hat{\mu}(\lambda\zeta) = \frac{(n-1)!(-1)^{n-1}}{2\pi i} \int\limits_{\{t \in \mathbb{C}: |t| = R\}} \frac{e^{-it}}{t^n} \Psi\left(\frac{\lambda\zeta}{t}\right) dt. \tag{2.17}$$

Due to (2.13) for the function $\Psi\left(\frac{\lambda\zeta}{t}\right)$ and $|t| = N + 1$ we have inequalities

$$\begin{aligned} \Psi\left(\frac{\lambda\zeta}{t}\right) &= \left[(\gamma'(\zeta))^{-1}\left(\frac{\lambda}{t} + d(\zeta)\frac{\lambda^2}{t^2}\right) + O_{K,\zeta}\left(\frac{\lambda^3}{t^3}\right)\right]^{N+1} \\ &= \left(\frac{\lambda}{\gamma'(\zeta) \cdot t}\right)^{N+1} \left[1 + d(\zeta)\frac{\lambda}{t} + O_{K,\zeta}\left(\frac{\lambda}{N+1}\right)^2\right]^{N+1} \end{aligned}$$

$$= \left(\frac{\lambda}{\gamma'(\zeta)\cdot t}\right)^{N+1}\Bigg[\left(1+d(\zeta)\frac{\lambda}{t}\right)^{N+1} + O_{K,\zeta}\left(\frac{|\lambda|^2}{N+1}\right)\Bigg]. \tag{2.18}$$

Substituting (2.18) into (2.17) we have

$$\hat{\mu}(\lambda\zeta) = (n-1)!(-1)^{n-1}[\mathcal{J}_1 + \mathcal{J}_2], \tag{2.19}$$

where

$$\mathcal{J}_1 = \frac{1}{2\pi i}\int\limits_{|t|=R} \frac{e^{-it}}{t^n}\left(\frac{\lambda}{\gamma'(\zeta)\cdot t}\right)^{N+1}\left(1+d(\zeta)\frac{\lambda}{t}\right)^{N+1} dt$$

$$\mathcal{J}_2 = \frac{1}{2\pi i}\int\limits_{|t|=N+1} \frac{e^{-it}}{t^n}\left(\frac{\lambda}{\gamma'(\zeta)\cdot t}\right)^{N+1} O_{K,\zeta}\left(\frac{|\lambda|^2}{N+1}\right) dt$$

Computing exactly $\mathcal{J}_1$ and estimating $\mathcal{J}_2$ we find

$$\mathcal{J}_1 = \frac{((\gamma')^{-1}\lambda)^{N+1}(-i)^{N+n}}{(N+n)!}\left[\sum_{\nu=0}^{N+1}\frac{(-id\cdot\lambda)^\nu}{\nu!}\prod_{j=1}^{\nu}\left(1-\frac{n+\nu-1}{N+n+j}\right)\right]$$

$$|\mathcal{J}_2| = \frac{1}{2\pi}\frac{((\gamma')^{-1}\cdot|\lambda|\cdot e)^{N+1}}{(N+1)^{n+N}}O_{K,\zeta}\left(\frac{|\lambda|^2}{N+1}\right). \tag{2.20}$$

It follows from (2.19), (2.20) that

$$|\hat{\mu}(\lambda\zeta)| \geq \frac{(n-1)!(|\gamma'(\zeta)|^{-1}|\lambda|)^{N+1}}{(N+n)!}\times \left[\left(1-O_{K,\zeta}\left(\frac{|\lambda|^2}{\sqrt{N+1}}\right)\right)e^{-|d(\zeta)|\cdot|\lambda|} - \frac{(N+n)!e^{N+1}}{(N+1)^{n+N}(n-1)!}O_{K,\zeta}\left(\frac{|\lambda|^2}{N+1}\right)\right]$$

$$= \frac{(n-1)!(|\gamma'(\zeta)|^{-1}\cdot|\lambda|)^{N+1}}{(N+n)!}$$

$$\left[\left(1 - O_{K,\zeta}\left(\frac{|\lambda|^2}{\sqrt{N+1}}\right)\right)e^{-|d(\zeta)||\lambda|} - O_{K,\zeta}\left(\frac{|\lambda|^2}{\sqrt{N+1}}\right)\right]$$

The estimate (1.12), and consequently, the theorem is proved.

References

[1] Natterer, F., *The mathematics of computerized tomography*. B. G. Teubner, Stuttgart, John Wiley and Sons, 1986.

[2] Zalcman, L., *Some inverse problems of potential theory*, Contemp. Math 6 (1987), 337–350.

[3] Tchakaloff, V., *Formules de cubatures mécaniques a coefficients non negatifs*, Bull. Sci. Math., Ser. 2, 1957, v. 81, N3, 123–134.

[4] Henkin, G., Shananin, A., *Bernstein theorems and Radon transform, Application to the theory of production functions*. Trans. of Math. monographs., 1990, v.81, 189–223.

[5] Akhiezer, N. I., *The classical moment problem and some related questions in analysis*, Hafner, New York, 1965.

[6] Siciak, J., *On some extremal functions and their application in the theory of analytic functions of several complex variables*. Trans. Am. Math. Soc. 1962, 105, 322–350.

[7] Zaharjuta, V. P., *Transfinite diameter, Cebysev constants and capacity for compact in* $\mathbb{C}^n$, Math. USSR Sbornik, 1975, 25, 350–364.

[8] Siciak, J., *An extremal problem in a class of plurisubharmonic functions*, Bull. Acad. Pol. Sci., 1976, 24, 563–568.

[9] Zaharjuta, V. P., *Extremal plurisubharmonic functions, orthonormal polynomials and the Bernstein-Walsh theorem for analytic functions of several complex variables*, Ann. Pol. Math. 1976/77, 33, 137–148.

[10] Stoll, W., *The characterization of strictly parabolic manifolds*. Ann. Scuola Norm. St. Pisa, 1980, 7, 87–154.

[11] Lempert, L., *La métrique de Kobayashi et la représentation des domaines sur la boule*, Bull. Soc. Mat. France 1981, 109, 427–474.

[12] ———, *Intrinsic distances and holomorphic retracts*. Proc. Conf. Varna, 1981, Complex Analysis and Applications, 81, Sofia, 1984, 341–364.

[13] Siciak, J., *Extremal plurisubharmonic functions in* $\mathbb{C}^n$, Ann. Polon. Math., 1981, 39, 175–211.

[14] Bedford, E., Taylor, B. A., *A new capacity for plurisubharmonic functions*, Acta Math., 1982, 149, 1–40.

[15] Lempert, L., *Solving the degenerate complex Monge-Ampère equation with one concentrated singularity*, Math. Ann., 1983, 263, 515–532.

[16] Levenberg, L., Taylor, B. A., *Comparison of capacities in* $\mathbb{C}^n$, Lecture Notes Springer, 1094, 1984, 162–171.

[17] Lempert, L., *Symmetries and other transformations of the complex Monge-Ampère equation*, Duke Math. Journal, 1985, v.52, 4, 869–885.

[18] Klimek, M., *Extremal plurisubharmonic functions and invariant pseudodistances*. Bull Soc. Math. France, 1985, 113, 123–142.

[19] Lundin, M., *The extremal plurisubharmonic functions for convex symmetric subsets of* R^n, Michigan Math. J., 1985, 32, 197–201.

[20] Demailly, J. P., *Mesures de Monge-Ampère et mesures plurisubharmoniques*, Math., Z., 1987, 194, 519–564.

[21] Lelong, P., *Notions capacitaires et fonctions de Green pluricomplexes dans les espaces de Banach*, C. R. Acad. Sci., Paris, 1987, 305, Série J, 71–76.

[22] Bedford, E., Taylor, B. A., *Plurisubharmonic functions with logarithmic singularities*, Ann. Inst. Fourier, Grenoble, 1988, 38, 4, 133–171.

[23] Martineau, A., *Sur la topologie des espaces de fonctions holomorphes*, Math. Ann., 1966, 163, 62–88.

[24] ———, *Equations différentielles d'ordre infini*, Bull. Soc. Math. France, 1967, 95, 109–154.

[25] Aizenberg, L. A., *Linear convexity in* $\mathbb{C}^n$ *and the distribution of the singularities of holomorphic functions*, Bull., Acad. Sci. Math., 1967, 15, 487–495.

[26] Gindikin, S. G., Henkin, G. M., *Integral geometry for* $\bar{\partial}$*-cohomology in* q *- linear concave domains in* $\mathbb{C}p^n$. Funct. Anal. Appl. 1979, 12, 247–261.

[27] Lelong, P., Gruman, L., *Entire functions of several complex variables*. Grundlehren, Math. Wiss., N282, Springer, 1986.

[28] Berndtsson, B., *A formula for interpolation and division in* $\mathbb{C}^n$, Math. Ann., 1983, 263, N4, 399–418.

[29] Henkin, G. M., *Method of integral representation in complex analysis, in Several Complex Variables* I, (Encycl. Math. Sc., 7), Springer-Verlag, 1990, 19–116.

NEVANLINNA THEOREMS IN PUSH-FORWARD VERSION

Shanyu Ji

I. Introduction

We consider a polynomial map f of $\mathbb{C}^n$, i.e., a holomporphic map $f : \mathbb{C}^n_z \to \mathbb{C}^n_w, z = (z_1, \ldots, z_n) \mapsto (f_1(z), \ldots, f_n(z))$, where $\mathbb{C}^n_z = \mathbb{C}^n_w = \mathbb{C}^n, z = (z_1, \ldots, z_n)$ and $w = (w_1, \ldots, w_n)$ are the coordinate systems for $\mathbb{C}^n_z$ and $\mathbb{C}^n_w$, respectively, and $f_1, \ldots, f_n \in \mathbb{C}[z_1, \ldots, z_n]$. For any polynomial map $f : \mathbb{C}^n_z \to \mathbb{C}^n_w$ with $\det(Df) \neq 0$, it is naturally associated a dominant rational map $F : \mathbb{P}^n_z - - \to \mathbb{P}^n_w$ defined by $[z_0 : z_1 : \ldots : z_n] \mapsto [z_0^{\deg f} : F_1(z_0, z_1, \ldots, z_n) : \ldots : F_n(z_0, \ldots, z_n)]$, where F_i is the homogeneous polynomial of degree $\deg f := \max_{1 \leq t \leq n} \deg f_t$ uniquely determined by $F_i(1, z_1, \ldots, z_n) = f_i(z_1, \ldots, z_n)$ for $i = 1, 2, \ldots, n$.

There is the well-known Jacobian problem which was raised by Keller in 1939 [**K**] and is still unknown (cf.[**V**]): If $f : \mathbb{C}^n_z \to \mathbb{C}^n_w$ is a polynomial map with the Jacobian $\det(Df) = 1$, then f has an inverse of polynomial map. In [**J**, corollary 4], we have proved: Let $f : \mathbb{C}^n_z \to \mathbb{C}^n_w$ be a polynomial map with $\det(Df) = 1$. Then f has an inverse of polynomial map if and only if $\text{supp}F_*D_{z_0} = \text{supp}D_{w_0}$, where D_{z_0} is the divisor given by $z_0 = 0$ and $F_*D_{z_0}$ is the push-forward current which is indeed a divisor.

From the above result, it leads us to take attention to the push-forward divisor $F_*D_{z_0}$. In order to investigate general push-forward divisors, in this paper, we establish the Nevanlinna main theorems in push-forward version which are analogous to the ones in the value distribution theory. We shall study any polynomial map $f : \mathbb{C}^n_z \to \mathbb{C}^n_w$ with $\det Df \neq 0$, its associated a dominant rational map $F : \mathbb{P}^n_z - - \to \mathbb{P}^n_w$ and any divisor D on $\mathbb{P}^n_z$. We shall prove the first main theorem, the second main theorem, the defect relation and some other results. For proving these theorems, besides the modified traditional method in the value distribution theory, some estimate from [**J**, theorem 2] about push-forward currents will be used.

This work is a part of the author's thesis. The author would like to thank his advisor Professor Shiffman for assistance and encouragement about this work. While preparing the final version of this work, the author is partially supported by a University of Houston Research Initiation Grant and by the NSF DMS-8922760.

2. Preliminaries

Meromorphic maps Let M and N be connected complex manifolds and let S be a proper analytic subset of M. Let $f : M - S \to N$ be a holomorphic map. The closed graph G of f is the closure of the graph of f over $M - S$ in $M \times N$. Let $\pi : G \to M$ and $\widehat{f} : G \to N$ be the natural projections. The map is said to be *meromorphic* on M, denoted by $f : M - - \to N$, if G is analytic in $M \times N$ and if π is proper. The *indeterminacy* $I_f = \{x \in M \mid \#\widehat{f}(\pi^{-1}(x)) > 1\}$ is analytic, where $\#B$ means the cardinacy of a set B, and is contained in S. We know $\operatorname{codim} I_f \geq 2$. We assume $S = I_f$.

If $B \subset M$ is a subset, we define the *image of* B *by* f is the set

$$f(B) = \widehat{f}(\pi^{-1}(B)) = \{y \in N \mid (x, y) \in G, \text{ for some } x \in B\}.$$

Currents on complex spaces Let X be a reduced, pure n-dimensional complex space and $\mathrm{Reg}(X)$ be the set of all regular points of X. We can define currents on X (cf.[**D**, p.14]). Since the problem is local, we assume that there is an embedding $j : X \to \Omega$, where $\Omega \subset \mathbb{C}^N$ is an open subset (i.e., X is identified with a closed analytic subset of Ω). We define

$$\mathcal{E}^{p,q}(X) = \tilde{j}^* \mathcal{E}^{p,q}(\Omega)$$

with the quotient topology, where

$$\tilde{j}^* : \mathcal{E}^{p,q}(\Omega) \to \mathcal{E}^{p,q}(\mathrm{Reg}(X))$$

is the usual pull-back map and $\mathcal{E}^{p,q}(M)$ is the space of all (p, q)-forms on a manifold M. It is known that the definition of $\mathcal{E}^{p,q}$ is indepedent of the choice of the embedding j (see [**D**, p.14]). Then we define $\mathcal{D}^{p,q}(X) = \{\sigma \in \mathcal{E}^{p,T}(X) \mid \sigma \textit{ has compact support on } X\}$ with the inductive limit topology. The dual space $\mathcal{D}'^{p,q}(X)$ is defined as the space of (p, q) bidimensional *currents* on X.

The operators $d, \partial, \bar{\partial}$ and push-forward of currents (by proper holomorphic maps) then are defined for currents on such complex spaces as defined on manifolds.

Divisors on complex spaces For any reduced, pure n-dimensional complex space $(X, \mathcal{O})$, we denote $\mathcal{M}$ the sheaf of germs of meromorphic functions on X and denote $\mathcal{M}^*$ the sheaf (of multiplication groups) of invertible elements in $\mathcal{M}$. Similary $\mathcal{O}^*$ is the sheaf of invertible elements in $\mathcal{O}$. A *divisor* D on X is a global section of the sheaf $\mathcal{M}^*/\mathcal{O}^*$. A divisor D on X also can be described by giving an open cover U_i of X and for each i an element $f_i \in \Gamma(U_i, \mathcal{M}^*)$ such that $f_i/f_j \in \Gamma(U_i \cap U_j, \mathcal{O}^*)$ for any i and j.

If $F : \mathbb{P}^n_z \to \mathbb{P}^n_w$ is a dominant rational (also meromorphic) map. Let G be the closed graph and $\pi, \widehat{F}$ are the natural projections. For any divisor D on $\mathbb{P}^n_z$, we can pull back it on G as a divisor $\pi^* D$ in an obvious way.

The Carlson-Griffiths singular form Let $D_1, \ldots, D_q$ be divisors on $\mathbb{P}^n_z$ such that $\mathrm{supp} D_1, \ldots, \mathrm{supp} D_q$ are manifolds located in normal crossings and each $D_j = D_{g_j}$, where $g_j \in \mathbb{C}[z_0, \ldots, z_n]$ is a homogeneous polynomial of degree p_j. Denote $D = \sum_{j=1}^q D_j$. Each D_j is also given by the system $\{U_i, g_j/z_i^{p_j}\}_{0 \le i \le n}$, where $U_i = \{[z_0 : \ldots : z_n] \in \mathbb{P}^n \mid z_i \neq 0\}$. The associated holomorphic line bundle L_{D_j} of D_j has the Hermitian metric $h_j = \{U_i, h_{ji}\}_{0 \le i \le n}$, where

$$h_{ji} = \frac{|z_i|^{2p_j}}{(|z_0|^2 + \ldots + |z_n|^2)^{p_j}}.$$

Let $L = L_{D_1} \otimes \ldots \otimes L_{D_q}$. Then the Hermitian metric h of L is $h = \{U_i, h_{1i} \cdot \ldots \cdot h_{qi}\}_{0 \le i \le n}$. For each D_j, as the section $\{g_j/z_i^{p_j}\}_{0 \le i \le n}$ of L_{D_j}, a globally defined function $\|D_j\|^2$ on $\mathbb{P}^n_z$ is defined by

$$\|D_j\|^2 \mid U_i = \frac{h_{ji}|g_j|^2}{eC_{g_j}|z_i|^{2p_j}} = \frac{|g_j|^2}{eC_{g_j}(|z_0|^2 + \ldots + |z_n|^2)^{p_j}},$$

where $C_{g_j} > 0$ is a constant such that

$$|g_j|^2 \leq C_{g_j}(|z_0|^2 + \ldots + |z_n|^2)^{p_j}$$

for all $z_0, \ldots, z_n \in \mathbb{C}$. We know the Chern form $c(K_{\mathbb{P}_z^n})$ of the canonical bundle $K_{\mathbb{P}_z^n}$

$$c(K_{\mathbb{P}_z^n}) = -(n+1)\Omega_{\mathbb{P}_z^n}, \tag{2.1}$$

which is also defined as the Ricci form of the volume form $\Omega_{\mathbb{P}_z^n}$ on $\mathbb{P}_z^n$. Let's recall the notions of Ricci form and volume form. Let M be any complex manifold. The canonical bundle K_M of M is the holomorphic line bundle whose transition functions are the Jacobian of the coordinate change mappings in the intersection of domains in a covering of M, i.e., let $\{U_\alpha, W^\alpha\}_\alpha$ be a coordinate syatem covering of M, then on $U_\alpha \cap U_\beta$, the transition functions $g_{\alpha\beta} = \det(\partial w_j^\alpha / \partial w_k^\beta)$. If $\Phi_\alpha = \prod_{v=1}^n \frac{\sqrt{-1}}{2\pi} dw_v^\alpha \wedge d\overline{w}_v^\alpha$ on U_α is the local Euclidean volume form, then a positive (n,n)-form Ω which is defined locally on U_α as $\lambda_\alpha \Phi_\alpha$ is a global form on M if and only if $\lambda_\beta = |g_{\alpha\beta}|^2 \lambda_\alpha$ in $U_\alpha \cap U_\beta$. Such (n,n)-form Ω is called a *volume form*. The *Ricci form* of Ω, denoted by $\operatorname{Ric}\Omega$, is defined by $\operatorname{Ric}\Omega \mid U_\alpha = dd^c \log \lambda_\alpha$.

The *Carlson-Griffiths Singular volume form* Ψ on $\mathbb{P}_z^n - \operatorname{supp} D$ is defined by (cf. [**SHA**, p.79])

$$\Psi = \frac{C\Omega_{\mathbb{P}_z^n}}{\prod_{j=1}^q (\log \|D_j\|^2)^2 \|D_j\|^2},$$

where the constant $C > 0$ is determined by the following properties

(2.2) $\operatorname{Ric}\Psi > 0$;

(2.3) $(\operatorname{Ric}\Psi)^n > \Psi$;

(2.4) $\int_{\mathbb{P}_z^n - \operatorname{supp} D} (\operatorname{Ric}\Psi)^n < +\infty$;

(2.5) $\operatorname{Ric}\Psi \mid (\mathbb{P}_z^n - \operatorname{supp} D) = c(L_D) + c(K_{\mathbb{P}_z^n})$
$\qquad - \sum_{j=1}^q dd^c \log(\|D_j\|^2)^2$.

3. Push-forward of currents by F

Let $f : \mathbb{C}_z^n \to \mathbb{C}_w^n$ be a polynomial map with the Jacobian $\det(Df) \neq 0$. Let $F : \mathbb{P}_z^n - - \to \mathbb{P}_w^n$ be its associated dominant rational map. Let G be the closed graph of F, and π and $\widehat{F}$ be the projections. G is an irreducible, pure complex n-dimensional analytic subset in $\mathbb{P}^n_z \times \mathbb{P}^n_w$, so G is regarded as an irreducible reduced complex space. Therefore π and $\widehat{F}$ are proper holomorphic maps from complex space onto complex manifolds.

For any divisor D on $\mathbb{P}^n_z$, we pull back it on G as a divisor $\pi^* D$. Then we obtain a pushforward current $\widehat{F}_*(\pi^* D)$ on $\mathbb{P}^n_w$. We want to show that this push-forward current is indeed a divisor. Before doing that, we need the following lemma. The proof below is due to Shiffman.

Lemma 3.1 *Let M and N be n-dimensional complex manifolds and let $f : M \to N$ be a surjective proper holomorphic map. If D is a divisor on M, then the current $f_* D$ is a divisor on N.*

Proof Let $A = \text{supp} D$, and $\bar{f} = f \mid A : A \to N$. We assume that $\text{codim}\, \bar{f}(A) = 1$. Let $S = \{x \in A \mid \dim \bar{f}^{-1}(\bar{f}(x)) \geq 1\}$. Then because of $\text{codim}\, \bar{f}(A) = 1$,

$$\text{codim}\, \bar{f}(S) \geq 2.$$

We first show that $f_* D \mid N - \bar{f}(S) \in \mathcal{D}'^{1,1}(n - \bar{f}(S))$ is a divisor. In fact, for any point $w \in N - \bar{f}(S), \bar{f}^{-1}(w)$ is a finite set. Then there is an open neighborhood $W(w)$ of w in $N - \bar{f}(S)$, and finite disjoint open subsets $U_1(w), \ldots, U_r(w)$ in M such that for each $U_i(w)$, there is a holomorphic function $g_i \in \mathcal{O}(U_i(w))$,

$$f^{-1}(W(w)) \cap A = \cup_{i=1}^r U_i(w) \cap A, \text{ and}$$
$$D \mid U_i(w) = dd^c \log |g_i|^2, \text{ for } i = 1, 2, \ldots, r,$$

where the Poincaré-Lelong formula is used. Let $J_f = \{z \in M \mid z \textit{ is a critical point of } f\}$. Then $f(J_f) \subset N$ is an analytic subset. Since $W(w) - f(J_f)$ is connected, there are integers $\lambda_1, \ldots, \lambda_r$ such that for any $u \in W(w) - f(J_f)$, there is an open neighborhood $W(u)$ of u in $W(w) - f(J_f)$, and disjoint open subsets $U_{1,1}(u), \ldots, U_{1,\lambda_1}(u) \subset U_1(w); \ldots; U_{r,1}(u), \ldots, U_{r,\lambda_r}(u) \subset U_r(w)$, so that $f \mid U_{i,j}(u) : U_{i,j}(u) \to W(u)$ is biholomorphic for all $i = 1, 2, \ldots;\ 1 \leq j \leq \lambda_i$. Then

$$
\begin{aligned}
f_*D \mid W(u) &= \sum_{i=1}^{r}\sum^{\lambda_i} dd^c \log \left|g_i \circ (f \mid U_{i,j}(u))^{-1}\right|^2 \\
&= dd^c \log \left| \prod_{i=1}^{r}\prod_{j=1}^{\lambda_i} g_i \circ (f \mid U_{i,j}(u))^{-1} \right|^2 \\
&= dd^c \log |g|^2 ,
\end{aligned}
$$

where $g = \prod_{i=j}^{r} \prod_{j=1}^{\lambda_i} g_i \circ (f|U_{i,j}(u))^{-1}$ on $W(u)$. g is a well-defined holomorphic function on $W(w) - f(J_f)$, which can be extended on $W(w)$ holomorphically. Therefore we have proved that f_*D is a divisor on $N - \bar{f}(S)$.

Let $V = \text{supp} f_*D \mid N - \bar{f}(S)$. V has a decomposition $V = \cup_j V_j$, where V_j are irreducible hypersurfaces on $N - f(S)$. Since we have proved $f_*D \mid N - f(S) = \sum_j n_j V_j$, V_j has an extension $\widetilde{V}_j$ in N for all j.

It suffices to show the current $T = \sum_j n_j \widetilde{V}_j - f_*D \in \mathcal{D}'^{1,1}(N)$ must be zero. Since $T \mid N - \bar{f}(S) = 0$, and $\dim_{\mathbb{R}} \bar{f}(S) \leq 2n - 4$. Then the current $T = 0$ follows from the following lemma. **QED**

Lemma 3.2 *Let* $0 < p < 2n$, *and* $\Omega \subset \mathbb{C}^n$ *be and open subset and* $E \subset \Omega$ *be a closed subset with* $h^p(E) = 0$, *where* h^p *is the Hausdorff measure of order* p. *If* $\sigma \in \mathcal{D}'^p(\Omega)$ *is d-closed and of order* 0, *then* $\|\sigma\|(E) = 0$.

Proof This is a special case of Federer [**F**, 4.1.20], or cf. [**Sh**, lemma A.2]. **QED**

Now we can prove that the current $\widehat{F}_*(\pi^* D)$ is a divisor. In fact, by Hironaka's theorem of resolution of singularities [**H**], there is a modification $\sigma : G' \to G$, where G' is a compact complex manifold. It follows that $\widehat{F}_* D = (\widehat{F} \circ \sigma)_*(\sigma^* D)$. Thus $\widehat{F}_* D$ is a divisor on $\mathbb{P}^n_w$ by the lemma above.

In [**J**], we proved that let $f : M - - \to N$ be a surjective meromorphic map, where M and N are compact connected complex manifolds of complex n-dimension. Let $\mathcal{L}$ be a semi-positive holomorphic line bundle over M with a nozero holomorphic section s. The locus of s on M is denoted by V as an analytic hypersurface. Then the image $f(V)$ is also an analytic hypersurface on N.

We take an open covering $\{U_\alpha\}$ of M and a Hermitian metric $h = \{h_\alpha\}$ of $\mathcal{L}$ such that the curvature of $(\mathcal{L}, h)$ is semi-positive. Let the given holomorphic section $s = \{s_\alpha\}$. Then we have a globally defined function on $M : \|s\|^2 = h_\alpha |s_\alpha|^2$ on U_α. Put $\varphi = -\log \|s\|^2$. By [**J**], $f_*\varphi$ is the plurisubharmonic exhaustion function of $N - f(V)$. By the lemma 3.1, if we denote D_s to be the divisor determined by s, $f_*D_s = \tilde{f}_*(\pi^* D_s)$ is also a divisor. Then for any point $w \in N \cap F(V)$, there exists an open neighborhood U_1 of w in N and a holomorphic function $g \in \mathcal{O}(U_1)$ such that $f_*D_s = dd^c \log |g|^2$. We notice that $f_*\varphi \in C^\infty(N - f(V \cup J_f \cup I_f))$. Then we can present

lemma 3.3 (See [**J**, theorem 2]) *Let $f, M, N, \mathcal{L}, s$ and φ be as above. Let w be any given point in $N \cap f(V)$ with a neighborhood U_1 as above. Then there exists an open neighborhood U of w with $U \subset U_1$ and a positive constant number $C = C(w, f, g)$ such that*

$$0 \leq f_*\varphi(u) \leq -\log |g(u)|^2 + C$$

for all $u \in U - f(V \cup J_f \cap I_f)$.

We can apply this theorem to any dominant rational map $F : \mathbb{P}^n_z - - \rightarrow \mathbb{P}^n_w$ and any holomorphic section s because of the fact that any hypersurface V on $\mathbb{P}^n_z$ should be a locus of some holomorphic section of some positive holomorphic line bundle over $\mathbb{P}^n_z$. For the section s, it is associated a globally defined function $\varphi = \|s\|$ on $\mathbb{P}^n_z$. It was proved that $F_*\varphi$ is an exhaustion plurisubharmonic function for $\mathbb{P}^n_w - F(V)$. Furthermore, the lemma 3.3 said that

$$F_*\varphi \in \mathcal{L}^1_{loc}(\mathbb{P}^n_w) \quad . \tag{3.4}$$

4. Notations in the value distribution theory

Let f and F be as before. Assume $\deg f > 1$. Consider the inclusion map $i : \mathbb{C}^n_w \hookrightarrow \mathbb{P}^n_w$, $(w_1, \ldots, w_n) \mapsto [1 : w_1 : \ldots : w_n]$, which identifies $i(\mathbb{C}^n_w) \cong \mathbb{C}^n_w$. We use $(w_1, \ldots, w_n)$ as coordinates system on $i(\mathbb{C}^n_w)$.

On $i(\mathbb{C}_w^n)$, we let

$$\begin{gathered}\varphi = dd^c(|w_1|^2 + \ldots + |w_n|^2), \quad \omega = dd^c \log(|w_1|^2 + \ldots + |w_n|^2), \\ \sigma_P = d^c \log(|w_1|^2 + \ldots + |w_n|^2) \wedge \omega^P; \quad \omega = \omega_{n-1}, \\ B(r) = \left\{[1 : w_1 : \ldots : w_n] \in i(\mathbb{C}_w^n) \mid |w_1|^2 + \ldots + |w_n|^2 < r^2\right\}, \\ S(r) = \left\{[1 : w_1 : \ldots : w_n] \in i(\mathbb{C}_w^n) \mid |w_1|^2 + \ldots + |w_n|^2 = r^2\right\}.\end{gathered}$$

Lemma 4.1 (Jenson-Lelong formula) *Let T be a real valued function and $T \in \mathcal{L}^1_{loc}(\mathbb{C}^n)$ such that dd^cT is of order 0. Then for $0 < r_0 < r$, one has*

$$\int_{R_0}^{r} \frac{dt}{t} \int_{B(t)} dd_cT \wedge \Omega^{n-1} = \frac{1}{2}\int_{S(r)} T \wedge \sigma_{n-1} - \frac{1}{2}\int_{S(r_0)} T \wedge \sigma_{n-1} + C,$$

where the constant C is independent of r.

Proof See [**Sh**, lemma 2.3]. **QED**

We define the *characteristic function of F* by

$$T_{*F}(r, r_0) = \int_{r_0}^{r} \frac{dt}{t} \int_{B(t)} F_*\Omega_{\mathbb{P}_z^n} \wedge \Omega^{n-1},$$

where $\Omega_{\mathbb{P}_z^n}$ is the Fubini-Study metric form on $\mathbb{P}_z^n$, and $F_* = \widehat{F}_*\pi^*$.

For any positive current χ on $i(\mathbb{C}_w^n)$ of $(n-1, n-1)$ bidimension, we define the *counting function* of χ by

$$N(\chi; r, r_0) = \int_{r_0}^{r} \frac{dt}{t^{2n-1}} \int_{B(t)} \chi \wedge \Omega^{n-1}.$$

Note that if χ is d-closed, by Stokes' theorem,

$$N(\chi; r, r_0) = \int_{r_0}^{r} \frac{dt}{t} \int_{B(t)} \chi \wedge \Omega^{n-1}.$$

Abbreviately, we denote

$$N_{*F}(D_g; r, r_0) = N(F_* D_g, r, r_0),$$

where D_g is a divisor on $\mathbb{P}_z^n$ given by a homogeneous polynomial g.

5. The first main theorem

Let $0 \neq g \in \mathbb{C}[z_0, z_1, \ldots, z_n]$ be any homogeneous polynomial. Denote D_g be its associated divisor on $\mathbb{P}_z^n$. Put

$$\varphi_g = \log \frac{eC_g(|z_0|^2 + \ldots + |z_n|^2)^{\deg g}}{|g(z_0, \ldots, z_n)|^2},$$

where C_g is a positive constant satisfying

$$|g(z_0, \ldots, z_n)|^2 \leq C_g(|z_0|^2 + \ldots + |z_n|^2)^{\deg g}$$

for all $z_0, \ldots, z_n \in \mathbb{C}$. Thus $\varphi_g \geq 0$ and $\varphi_g \in C^\infty(\mathbb{P}_z^n - \mathrm{supp} D_g) \cap \mathcal{L}_{loc}^1(\mathbb{P}_z^n)$.

Apply the Poincaré-Lelong formula, we see

$$dd^c \varphi_g = \deg g \cdot \Omega_{\mathbb{P}_z^n} - D_g, \qquad \text{on } \mathbb{P}_z^n.$$

Then

$$F_* dd^c \varphi_g = \deg g \cdot F_* \Omega_{\mathbb{P}_z^n} - F_* D_g, \qquad \text{on } \mathbb{P}_w^n.$$

Since $\widehat{F}_*, \pi^*$ commute with d, d^c, we see F_* commutes with dd^c, then we restrict the above relation $i(\mathbb{C}_w^n)$ to obtain

$$dd^c F_* \varphi_g = \deg g \cdot F_* \Omega_{\mathbb{P}_z^n} - F_* D_g, \qquad \text{on } i(\mathbb{P}_w^n).$$

By (3.4), it follows that

$$F_* \varphi_g \in \mathcal{L}_{loc}^1(\mathbb{C}_w^n).$$

Then by applying the lemma 4.1, we have proved

Theorem 5.1 (First main theorem) *Let f, F be as before. Then*

$$\deg g \cdot T_{*F}(r, r_0) = N_{*F}(D_g; r, r_0) + \frac{1}{2} \int\limits_{S(r)} F_* \varphi_g \sigma + O(1)$$

for $r \gg r_0$.

Corollary 5.2 (Nevanlinna inequality)

$$N_{*F}(D_g; r, r_0) \leq \deg g T_{*F}(r, r_0) + O(1).$$

Proof Note $\varphi_g \geq 0$, then $\int_{S(r)} F_* \varphi_g \sigma \leq 0$. **QED**

We would like to give the following proposition to close this section. The inequality here is conjectured to be equality which remains a problem.

Proposition 5.3 $\overline{\lim}_{r \to \infty} \frac{T_{*F}(r, r_0)}{\log r} \leq (\deg f)^{n-1}$.

Proof

$$\overline{\lim}_{r \to \infty} \frac{T_{*F}(r, r_0)}{\log r} \leq \overline{\lim}_{r \to \infty} \int\limits_{B(r)} F_* \omega_{\mathbb{P}_z^n} \wedge \omega^{n-1}$$

$$= \int\limits_{\mathbb{P}_z^n} F_* \omega_{\mathbb{P}_z^n} \wedge \omega^{n-1}$$

$$= \int\limits_{G} \pi^* \omega_{\mathbb{P}_z^n} \wedge \widehat{F}^* \omega^{n-1}$$

$$= \int\limits_{\mathbb{C}_z^n} dd^c \log(1 + |z_1|^2 + \ldots + |z_n|^2) \wedge \wedge f^*(dd^c \log(|w_1|^2 + \ldots + |w_n|^2))^{n-1}$$

$$= \lim{}_{r \to \infty} \left[\int_{r_0}^{r} \frac{dt}{t} \int\limits_{B(t, \mathbb{C}_z^n)} dd^c \log(1 + |z_1|^2 + \ldots + |z_n|^2) \wedge \wedge (dd^c \log(|f_1|^2 + \ldots + |f_n|^2))^{n-1} \right] / \log r,$$

where G is the closed graph of F, and $B(r, \mathbb{C}_z^n) = \{(z_1, \ldots, z_n) \in \mathbb{C}_z^n \mid |z_1|^2 + \ldots + |z_n|^2 < r^2\}$.

Since

$$\int_{r_0}^{r} \frac{dt}{t} \int_{B(t,\mathbb{C}_z^n)} dd^c \log(1 + |z_1|^2 + \ldots + |z_n|^2))^{n-1} \wedge (dd^c \log(|f_1|^2 + \ldots + |f_n|^2))^{n-1}$$

$$\leq \frac{1}{2} \int_{\partial B(r,\mathbb{C}_z^n)} \log |f|^2 (dd^c \log |f|^2)^{n-2} \wedge dd^c \log(1 + |z|^2) \wedge \wedge d^c \log(1 + |z|^2)$$

$$- \frac{1}{2} \int_{\partial B(r_0,\mathbb{C}_z^n)} \log |f|^2 (dd^c \log |f|^2)^{n-2} \wedge dd^c \log(1 + |z|^2) \wedge \wedge d^c \log(1 + |z|^2) + O(1)$$

$$\leq (\deg f \cdot \log r + \frac{A}{2}) \int_{\partial B(r,\mathbb{C}_z^n)} (dd^c \log |f|^2)^{n-2} \wedge dd^c \log(1 + |z|^2) \wedge \wedge d^c \log(1 + |z|^2) + O(1)$$

$$\leq (\deg f \cdot \log r + A) \int_{B(r,\mathbb{C}_z^n)} (dd^c \log |f|^2)^{n-2} \wedge (dd^c \log(1 + |z|^2))^2 + O(1),$$

where $|f|^2 = |f_1|^2 + \ldots + |f_n|^2$ and the positive constant A is independent of r, we then have

$$\overline{\lim}_{r\to\infty} \frac{T_{*F}(r, r_0)}{\log r}$$

$$\leq \lim_{r\to\infty} \frac{(\deg f \cdot \log r + A)}{\log r} \int_{B(r,\mathbb{C}_z^n)} (dd^c \log |f|^2)^{n-2} \wedge \wedge (dd^c \log(1 + |z|^2))^2$$

$$\leq \deg f \int_{\mathbb{C}_z^n} (dd^c \log |f|^2)^{n-2} \wedge (dd^c \log(1 + |z|^2))^2$$

$$
\begin{aligned}
&\leq \ldots\ldots \\
&\leq (\deg f)^{n-1} \int_{\mathbb{C}_z^n} (dd^c \log(1+|z|^2))^n \\
&= (\deg f)^{n-1}.
\end{aligned}
$$

QED

6. The second main theorem

In this section, we shall prove the second main theorem. Let f, F be as before. Let D_{J_F} be the *ramification divisor* of the meromorphic map F on $\mathbb{P}_z^n$, i.e., locally on $\mathbb{P}_z^n - I_F$, it is given by the Jacobian determinant of F, and then it is extended on $\mathbb{P}_z^n$ (cf.[**SH**, p.73]). Let D_{J_F} be determined by a unique (up to a constant factor) homogeneous polynomial $J_F \in \mathbb{C}[z_0, \ldots, z_n]$. We use λ_F to denote the sheets number of F.

Theorem 6.1 (Second main theorem) *Let f, F be as before. Let $D_1, \ldots, D_q$ be divisors on $\mathbb{P}_z^n$ so that* $\operatorname{supp}D_1, \ldots, \operatorname{supp}D_q$ *are manifolds located in normal crossings. Suppose each $D_j = D_{g_j}$, where $g_j \in \mathbb{C}[z_0, \ldots, z_n]$ is homogeneous polynomial of degree p_j, for $j = 1, 2, \ldots, q$. Denote $D = \sum_{j=1}^q D_j$. Then*

$$
\left(\sum_{j=1}^q p_j - (n+1)\right) T_{*F}(r, r_0) \leq N_{*F}(D; r, r_0) + N_{*F}(D_{J_F}; r, r_0) + \log r + O(1)
$$

for $r \gg r_0$.

Proof For any $w \in i(\mathbb{C}_w^n) - F(\operatorname{supp}D_{J_F})$, there is an open neighborhood $W(w)$ of w in $i(\mathbb{C}_w^n) - F(\operatorname{supp}D_{J_F})$ and λ_F disjoint open subsets $U_1(w), \ldots, U_{\lambda_F}(w)$ in $\mathbb{C}_z^n$ such that

$$
F^{-1}(W(w)) = \bigcup_{j=1}^{\lambda_f} U_j(w) \text{ and}
$$

the restriction $F \mid U_i(w)$ is biholomorphic.

Consider the Carlson-Griffiths singular volume form Ψ,

$$
\begin{aligned}
&F_*\Psi \mid W(w) \\
&= \sum_{i=1}^{\lambda_F}(F \mid U_i(w))_*(\Psi \mid U_i(w)) \\
&= \sum_{i=1}^{\lambda_F}(F \mid U_i(w))_* \frac{dd^c \log(1+|z_1|^2+\ldots+|z_n|^2))^n}{\prod_{j=1}^q(\log\|D_j\|^2)^2\|D_j\|^2} \\
&= \sum_{i=1}^{\lambda_F}\left((F \mid U_i(w))^{-1}\right)^* \frac{n!(dd^c(|z_1|^2+\ldots+|z_n|^2))^n}{(1+|z_1|^2+\ldots+|z_n|^2)^{n+1}} \times \\
&\qquad \times \frac{1}{\prod_{j=1}^q(\log\|D_j\|^2)^2\|D_j\|^2} \\
&= \tilde{\xi}(dd^c(|w_1|^2+\ldots+|w_n|^2))^n,
\end{aligned}
$$

where

$$
\tilde{\xi} = \sum_{i=1}^{\lambda_F} \frac{n!}{((F \mid U_i(w))^{-1})^* X(w)},
$$

and

$$
\begin{aligned}
X(w) = \Bigg[&| \det D(F \mid U_i(w)) \mid (1+\sum_{t=1}^n |z_t|^2)^{n+1} \times \\
&\times \prod_{j=1}^q (\log\|D_j\|^2)^2\|D_j\|^2\Bigg].
\end{aligned}
$$

Put

$$
\xi = \prod_{i=1}^{\lambda_F} \frac{n!}{((F \mid U_i(w))^{-1})^* X(w)}
$$

Apply the Poincaré-Lelong formula, $dd^c \log\|D_j\|^2 = -c(L_{D_j}) + D_j$, so

$$
\begin{aligned}
dd^c \log \xi &= F_*c(L_{D_j}) + F_*c(K_{\mathbb{P}_z^n}) \\
&\qquad - \sum_{j=1}^q F_* dd^c \log(\log\|D_j\|^2)^2 - F_*D - F_*D_{J_F} \\
&= F_* \operatorname{Ric}\Psi - F_*D - F_*D_{J_F},
\end{aligned}
$$

where the formula (2.5) was used.

By (2.1) we obtain

$$N(dd^c \log \xi; r, r_0) + \sum_{j=1}^{q} N(F_* dd^c \log(\log \|D_j\|^2)^2; r, r_0)$$
$$= \left(\sum_{j=1}^{q} p_j - (n+1)\right) T_{*F}(r, r_0)$$
$$- N_{*F}(D; r, r_0) - N_{*F}(D_{J_F}; r, r_0) \tag{6.2}$$

To prove the theorem, it suffices to estimate the left hand side of the above identity.

Let's estimate $N(F_* dd^c \log(\log \|D_j\|^2)^2; r, r_0)$ first. Since

$$\log(\log \|D_j\|^2)^2 = \log\left(\log \frac{|g_j(z_0, \ldots, z_n)|^2}{eC_{g_j}(|z_0|^2 + \ldots + |z_n|^2)^{\deg g_j})^2}\right)^2,$$

then

$$\varphi_{g_j} = \log \frac{eC_{g_j}(|z_0|^2 + \ldots + |z_n|^2)^{\deg g_j}}{|g_j(z_0, \ldots, z_n)|^2} \geq 1.$$

Thus

$$0 \leq F_* \log(\log \|D_j\|^2)^2 \leq 2F^* \log \varphi_{g_j} + 2\log 2. \tag{6.3}$$

Take any $w \in i(\mathbb{C}^n_w)$, take $W(w)$ and $U_1(w), \ldots, U_{\lambda_F}(w)$ as before, we then have

$$F^* \log \varphi_{g_j} \mid W(w) = \sum_{v=1}^{\lambda_F} (F \mid U_v(w))_* \log \varphi_{g_j}$$
$$= \log \prod_{v=1}^{\lambda_F} \varphi_{g_j} \circ (F \mid U_v(w))^{-1} \tag{6.4}$$
$$= \lambda_F \log \prod_{v=1}^{\lambda_F} \left(\varphi_{g_j} \circ (F \mid U_v(w))^{-1}\right)^{1/\lambda_F}$$

$$\leq \lambda_F \log \sum_{v=1}^{\lambda_F} \varphi_{g_j} \circ (F \mid U_v(w))^{-1} - \lambda_F \log \lambda_F. \tag{6.4}$$
$$= \lambda_F \log F_* \varphi_{g_j} - \lambda_F \log \lambda_F.$$

By the proof of the lemma 3.3, we also know

$$F_* \log(\log \|D_j\|^2)^2 \in \mathcal{L}^1_{loc}(i(\mathbb{C}^n)).$$

Thus apply the lemma 4.1 and by (6.3), (6.4), we obtain

$$\begin{aligned}
&\sum_{j=1}^{q} N(F_* dd^c \log(\log \|D_j\|^2)^2; r, r_0) \\
&\quad \leq -\frac{1}{2} \sum_{j=1}^{q} \int_{S(r)} F_* \log(\log \|D_j\|^2)^2 \sigma + O(1) \\
&\quad \leq \sum_{j=1}^{q} \int_{S(r)} (F_* \log \varphi_{g_j}) \sigma + O(1) \\
&\quad \leq \lambda_F \sum_{j=1}^{q} \int_{S(r)} (\log F_* \varphi_{g_j}) \sigma + O(1) \qquad (6.5) \\
&\quad \leq \lambda_F \sum_{j=1}^{q} \log \int_{S(r)} (F_* \varphi_{g_j}) \sigma + O(1) \\
&\quad \leq \lambda_F \sum_{j=1}^{q} \log(2 \deg g_j \, T_{*F}(r, r_0)) + O(1) \\
&\quad \leq O(\log T_{*F}(r, r_0))
\end{aligned}$$

for $r \gg r_0$. Here the last second inequality is due to the first main theorem 5.1.

Next we estimate the term $N(dd^c \log \xi; r, r_0)$ in (6.2). By the previous argument, we see $\log \xi \in \mathcal{L}^1_{loc}(i(\mathbb{C}^n_w))$. Then

$$
\begin{aligned}
N(dd^c \log \xi; r, r_0) &= \frac{1}{2} \int_{S(r)} \log \xi^{1/\lambda_F} \sigma + O(1) \\
&= \frac{\lambda_F}{2} \int_{S(r)} \log \xi^{1/\lambda_F} \sigma + O(1) \\
&\leq \frac{\lambda_F}{2} \int_{S(r)} \log \tilde{\xi} \sigma + O(1) \\
&\quad \textit{(By the definitions of } \xi \textit{ and } \tilde{\xi}, \\
&\quad \textit{and by } \xi^{1/\lambda_F} \leq \tilde{\xi}/\lambda_F) \\
&= \frac{\lambda_F}{2} \int_{S(r)} \log c \tilde{\xi}^{\frac{1}{n}} \sigma + O(1) \\
&\leq \frac{\lambda_F}{2} \log \int_{S(r)} c \tilde{\xi}^{\frac{1}{n}} \sigma + O(1),
\end{aligned}
\tag{6.6}
$$

where $c = \frac{1}{(n!)^{1/n}}$. Since

$$
\int_{B(t)} (c\tilde{\xi}^{\frac{1}{n}}) \varphi^n = 2 \int_0^r \left(\int_{S(t)} (c\tilde{\xi}^{\frac{1}{n}} \sigma) \right) t^{2n-1} dt,
$$

$$
\int_{S(r)} (c\tilde{\xi}^{\frac{1}{n}}) \sigma = \frac{1}{2r^{2n-1}} \frac{d}{dr} \int_{B(r)} (c\tilde{\xi}^{\frac{1}{n}}) \varphi^n. \tag{6.7}
$$

Put

$$
\widehat{T}(r, r_0) = \int_{r_0}^{r} \frac{dt}{t^{2n-1}} \int_{B(t)} (c\tilde{\xi}^{\frac{1}{n}}) \varphi^n. \tag{6.8}
$$

From (6.6), (6.7) and (6.8), it follows

$$
\begin{aligned}
&N(dd^c \log \xi; r, r_0) \\
&\qquad \leq \frac{\lambda_F n}{2} \log \left(\frac{1}{2nr^{2n-1}} \frac{d}{dr} \left(r^{2n-1} \frac{d\widehat{T}}{dr} \right) \right) + O(1).
\end{aligned}
\tag{6.9}
$$

By the classical result in the value distribution theory (cf.[**Sha**, p.84]), (6.9) implies that for any $\epsilon > 0$, there is $\delta = \delta(\epsilon) > 0$, so that $\delta(\epsilon) \to 0$ as $\epsilon \to 0$ and there is a subset $E = E(\epsilon) \subset \mathbb{R}_+$ with finite δ-measure, such that

$$
\begin{aligned}
&N(dd^c \log \xi; r, r_0) \\
&\qquad \leq \epsilon \log r + O\left(\log \widehat{T}(r, r_0) \right) + O(1),
\end{aligned}
\tag{6.10}
$$

for all $r \in \mathbb{R}_+ - E$.

To complete estimating (6.10), we estimate the term $\widehat{T}(r, r_0)$. Let $F_*\Psi \mid (i(\mathbb{C}^n_w) - F(\text{supp}D \cup \text{supp}D_{J_F})) = \sum_{j,k}^n R_{jk} dw_j \wedge d\overline{w_k}$, where the matrix $R = (R_{jk})$ is positive definite. Recall the definition of $\tilde{\xi}$ and (2.3),

$$
\begin{aligned}
&\tilde{\xi}(dd^c(|w_1|^2 + \ldots + |w_n|^2))^n \\
&\quad = F_*\Psi \\
&\quad \leq F_*(\text{Ric}\,\Psi)^n \\
&\quad = n! \det R (dd^c(|w_1|^2 + \ldots + |w_n|^2))^n.
\end{aligned}
$$

Thus $\tilde{\xi} \leq n! \det R$ holds on $i(\mathbb{C}^n_w) - F(\text{supp}D \cup \text{supp}D_{J_F})$. By Hardamard inequality: for any positive definity matrix R, $(\det R)^{1/n} \leq \frac{1}{n} \operatorname{tr} R$. Then we have

$$
\begin{aligned}
&\left(\frac{\tilde{\xi}}{n!} \right)^{1/n} (dd^c(|w_1|^2 + \ldots + |w_n|^2))^n \\
&\quad \leq \frac{1}{n} \operatorname{tr} R \left(dd^c(|w_1|^2 + \ldots + |w_n|^2) \right)^n \\
&\quad \leq \frac{1}{n} \sum_{j=1}^n R_{jj} \left(dd^c(|w_1|^2 + \ldots + |w_n|^2) \right)^n \\
&\quad = F_* \,\text{Ric}\,\Psi \wedge \left(dd^c(|w_1|^2 + \ldots + |w_n|^2) \right)^{n-1},
\end{aligned}
\tag{6.11}
$$

where since $\left(dd^c(|w_1|^2+\ldots+|w_n|^2)\right)^{n-1}=(n-1)!\sum_{j=1}^n\left(\frac{\sqrt{-1}}{2\pi}\right)^{n-1}$ $dw_1\wedge d\overline{w}_1\wedge\ldots(\wedge dw_j)^{\text{omit}}\wedge(d\overline{w_j})^{\text{omit}}\wedge\ldots\wedge dw_n\wedge d\overline{w_n}$, $F_*\operatorname{Ric}\Psi$ $\wedge\ \left(dd^c(|w_1|^2+\ldots+|w_n|^2)\right)^{n-1}\ =\ \frac{1}{n}\sum_{j=1}^n R_{jj}\left(dd^c(|w_1|^2+\ldots+|w_n|^2)\right)^n$ by direct computation.

Recall the definition of $\widehat{T}(r,r_0)$ and (6.4), we see

$$\begin{aligned}\widehat{T}(r,r_0)&=\int_{r_0}^{r}\frac{dt}{t^{2n-1}}\int_{B(t)}(c\tilde{\xi})^{1/n}\varphi^n\\&\le\int_{r_0}^{r}\frac{dt}{t^{2n-1}}\int_{B(t)}F_*\operatorname{Ric}\Psi\wedge\left(dd^c(|w_1|^2+\ldots+|w_n|^2)\right)^{n-1}\\&=N(F_*\operatorname{Ric}\Psi;r,r_0)\\&=\left(\sum_{j=1}^{q}p_j-(n+1)\right)T_{*F}(r,r_0)\\&\qquad\qquad+N\left(F_*dd^c\log(\log\|D_j\|^2)^2;r,r_0\right)\\&\le O(T_{*F}(r,r_0)).\end{aligned}$$

Together with (6.10), for any $\epsilon>0$, and $r\in\mathbb{R}_+-E$,

$$\begin{aligned}N(dd^c\log\xi;r,r_0)&\le\epsilon\log r+O(\log T_{*F}(r,r_0))+O(1)\\&\le\epsilon\log r+O(\log^+\log r)+O(1).\end{aligned}\qquad(6.13)$$

Here the proposition (5.3) is used. Also by a classical result (cf. [**Sha**, remark 1, p.88]), we obtain

$$N(dd^c\log\xi;r,r_0)\le\epsilon\log r+O(\log^+\log r)\qquad(6.13)$$

for all $r\gg r_0$.

Combinating (6.2), (6.5) and (6.13), we proved the theorem.

QED

7. Other results

For any divisor $D=D_g$ on $\mathbb{P}_z^n$, we define the *defect of* D *under* f by

$$\delta_{*F} = \deg g - \overline{\lim}_{r\to\infty} \frac{N_{*F}(D; r, r_0)}{T_{*F}(r, r_0)},$$

where $\delta_{*F}(D)$ is independent of the choice of r_0. By the second main theorem, we have the following

Theorem 7.1 (Defect relation) *Let $f, F, D_1, \ldots, D_q$ be as in the theorem 6.1. Then*

$$\sum_{j=1}^{q} \delta_{*F}(D_j) \le n + 1 + \deg(J_F).$$

Theorem 7.2 *Let f, F be as above. Let D be a divisor given by a hyperplane on $\mathbb{P}^n_w$ with* $\operatorname{codim}(\operatorname{supp} D \cap F(\operatorname{supp} D_{J_F})) \ge 2$. *Let $F^*D = D_g$, and suppose that* $\operatorname{supp} D_{g_1}$ *is smooth and g_1 divides g, where $g_1, g \in \mathbb{C}[z_0, \ldots, z_n]$ are homogeneous polynomials. Then*

$$\deg g_1 \le \lambda_F + n + 1 + \deg J_F.$$

More precisely,

$$\deg g_1 \le \lambda_F \underline{\lim}_{r\to\infty} \frac{\log r}{T_{*F}(r, r_0)} + n + 1 + \deg J_F.$$

Proof If we can show that

$$N_{*F}(D_{g_1}; r, r_0) \le \lambda_F N(D : r, r_0), \tag{7.3}$$

then by the theorem 6.1, for any $\epsilon > 0$ and $0 < r_0 < r < +\infty$,

$$\begin{aligned}
&(\deg g_1 - (n+1))T_{*F}(r, r_0) \\
&\quad \le N_{*F}(D_{g_1}; r, r_0) + N_{*F}(D_{J_F}; r, r_0) + \epsilon \log r + O(1) \\
&\quad \le \lambda_F N(D; r, r_0) + \deg J_F \cdot T_{*F}(r, r_0) + \epsilon \log r + O(1) \\
&\quad \le \lambda_F \log r + \deg J_F \cdot T_{*F}(r, r_0) + \epsilon \log + O(1).
\end{aligned}$$

Here the hypothesis that D is a hyperplane is used.

Now we prove (7.3). By the proof of [**Dr**, lemma 3.2], we know

$$F^*D \leq [\text{supp}F^*D] + D_{J_F}$$

holds on $i(\mathbb{C}_z^n)$, where $[\text{supp}F^*D]$ is the current by integration on $\text{supp}F^*D$. Then if $U \subset i(\mathbb{C}_z^n)$ is an open subset with $U \cap \text{supp}D_{J_P} = \emptyset$,

$$[\text{supp}F^*D] \mid U \leq F^*D \mid U \leq [\text{supp}F^*D] \mid U.$$

$$\text{i.e., } F^*D \mid U = [\text{supp}F^*D] \mid U. \tag{7.4}$$

For any $w \in i(\mathbb{C}_w^n) - F(\text{supp}D_{J_F})$. There is an open neighborhood $W(w)$ of w in $i(\mathbb{C}_w^n) - F(\text{supp}D_{J_F})$ and λ_F disjoint open subsets $U_1(w), \ldots, U_{\lambda_F}(w)$ in $i(\mathbb{C}_z^n)$, such that $F^{-1}(W(w)) = \bigcup_{i=1}^{\lambda_F} U_i(w)$, and $F \mid U_i(w) : U_i(w) \to W(w)$ is biholomorphic. Then

$$\begin{aligned} F_*D_{g_1} &\leq F_*D_g \mid W(w) = F_*[\text{supp}D_g] \mid W(w) \\ &= \sum_{i=1}^{\lambda_F} (F \mid U_i(w))^{-1*}[\text{supp}D_g] \\ &= \sum_{i=1}^{\lambda_F} \left[\text{supp}(F \mid U_i(w))^{-1*}[\text{supp}D_g]\right] \\ &= \sum_{i=1}^{\lambda_F} [\text{supp}D] \\ &= \lambda_F D. \end{aligned}$$

Thus $F_*D_g \mid i(\mathbb{C}_w^n) - F(\text{supp}D_{J_F}) = \lambda_F D$. Since $\text{codim}\,(\text{supp}D \cap F(\text{supp}D_{J_F})) \geq 2$, then $F_*D_g \mid i(\mathbb{C}_w^n) = \lambda_F D$, i.e.,

$$F_*D_{g_1} \leq F_*D_g = \lambda_F D$$

holds on $i(\mathbb{C}_w^n)$. This proves (7.3). **QED**

REFERENCES

[D] J.-P. Demailly, Measures de Monge-Ampère et caractérisation géométrique des variétes algébriques affines, *Mèm. Soc. Math. France (N.S.)* 19(1985), 1–124.

[Dr] S.J. Drouihet, A unicity theorem for meromorphic mappings between algebraically varieties, *Trans. Amer. Math. Soc.* 265(1981), 349–358.

[F] H. Federer, *Geometric measure theory*, Springer-Verlag, Berlin-Heidelberg-New York (1969).

[H] H. Hironaka, Resolution of singularities of an algebraic variety, I, II, *Ann. of Math.* 79(1964), 109–326.

[K] O.H. Keller, Ganze Cremona-Transformationen, *Monatshefte für Math. und Phys.*, 47(1939), 299–306.

[J] S. Ji, Image of analytic hypersurfaces, *Indiana Univ. of Math. J.*, Vol. 39, 2(1990), 477–483.

[Sha] B.V. Shabat, Distribution of values of holomorphic mappings, *Transl. of Math. Mono.* Vol. 61, A.M.S., 1985.

[Sh] B. Shiffman, Introduction to the Carlson-Griffiths equidistribution theory, *Lecture Notes in Math.* 981, Springer-Verlag, (1983) 44–89.

[St] W. Stoll, Introduction to value distribution theory of meromorphic maps, in *Complex analysis, Lecture Notes in Math.* 950, Springer-Verlag, (1982) 210–359.

[V] A.G. Vitushkin, On polynomial transformations of $\mathbb{C}^n$, in *Manifolds*, Tokyo Univ. Press, Tokyo (1975), 415–417.

RECENT WORK ON NEVANLINNA THEORY AND DIOPHANTINE APPROXIMATIONS

Paul Vojta

What I will describe here is a formal analogy between value distribution theory and various diophantine questions in number theory. In particular, there is a dictionary which can be used to translate, e. g., the First and Second Main Theorems of Nevanlinna theory into the number field case. For example, we shall see that the number theoretic counterpart to the Second Main Theorem combines Roth's theorem and Mordell's conjecture (proved by Faltings in 1983).

This analogy is only formal, though: it can only be used to translate the statements of main results, and the proofs of some of their corollaries. The proofs of the main results, though, cannot be translated due to a lack of a number theoretic analogue of the derivative of a meromorphic function, among other reasons. All that I can say at this point is that negative curvature plays a role in the proofs in both cases.

Thus, until recently the analogy was good only for producing conjectures, by translating statements of theorems in value distribution theory into number theory. But in 1989 it has played a role in finding a new proof of the Mordell conjecture, via the suggestion that the Mordell conjecture and Roth's theorem should have a common proof, as is the case with the Second Main Theorem.

We begin by briefly describing this analogy, but only briefly as it has been described elsewhere in [**V 1**] and [**V 2**], as well as in the book [**V 3**]. Likewise, more recent results will be described in [**V 6**]; therefore we refer the reader to [**V 3**] and [**V 6**] for details.

Let $f : \mathbb{C} \to C$ be a holomorphic curve in a compact Riemann surface (which we may assume is connected). Let D be an effective reduced divisor on C; i.e., a finite set of points, and let $\operatorname{dist}(D, P)$ be some function measuring the distance from P to a fixed divisor D. Then we have the usual definition

---

Partially supported by the National Science Foundation Grants DMS-8610730 and DMS-9001372.

$$m(D,r) = \int_0^{2\pi} -\log \operatorname{dist}(D, f(re^{i\theta}))\frac{d\theta}{2\pi}.$$

Assuming that $f(0) \notin \operatorname{Supp}D$, the definition of the counting function can be rewritten as

$$N(D,r) = \sum_{w \in \mathbb{D}_r} \operatorname{ord}_w f^*D \cdot \log \frac{r}{|w|}.$$

Finally, let the characteristic function be given by the more classical definition:

$$T_D(r) = m(D,r) + N(D,r).$$

Note in particular that in the above definitions, we only needed the restriction of f to the closed disc $\bar{\mathbb{D}}_r$, of radius r. Thus we are actually regarding f as an infinite family of maps $f_r : \bar{\mathbb{D}}_r \to C$, obtained by restriction from f. In the analogy with number theory, let each f_r correspond to one of (countably many) rational points, so that a holomorphic function $f : \mathbb{C} \to C$ corresponds to an *infinite* set of rational points on C. For example there are no infinite sets of (distinct) rational points on a curve of genus > 1 (Mordell's conjecture), just as there are no nontrivial holomorphic maps from $\mathbb{C}$ to a Riemann surface of genus > 1. Both these facts follow from the appropriate version of the Second Main Theorem, as defined below.

To make the number theoretic counterparts to the standard definitions as above, let C be a smooth connected projective curve, and let D be a reduced effective divisor on C. Assume that both C and D are defined over a number field k. For each place v of k (i.e., for each complex embedding $\sigma : k \to \mathbb{C}$ and for each non-archimedean absolute value corresponding to a prime ideal in the ring of integers of k), let $\operatorname{dist}_v(D, P)$ again be the distance from P to a fixed divisor D in the v-adic topology. These distances should be chosen consistently, as in ([**L 2**], Ch. 10, Sect. 2). For example, if $C = \mathbb{P}^1$ and $D = [\alpha]$, then the various $\operatorname{dist}_v([\alpha], P)$ functions can be written as $\min(1, |x - \alpha|_v)$.

Then the proximity function is defined as

$$m(D,P)=\frac{1}{[k:\mathbb{Q}]}\sum_{v|\infty}-\log\operatorname{dist}_v(D,P)$$

where the notation $v|\infty$ means the sum is taken over the (finitely many) archimedean places of k. Thus, we are comparing the absolute values of f on the boundary of $\mathbb{D}_r$, with the absolute values "at infinity" of a number field.

The formula for the counting function is similar:

$$N(D,P)=\frac{1}{[k:\mathbb{Q}]}\sum_{v\nmid\infty}-\log\operatorname{dist}_v(D,P).$$

This is more clearly a counterpart to the definition in the Nevanlinna case if we write it as

$$N(D,P)=\frac{1}{[k:\mathbb{Q}]}\sum_{v\nmid\infty}\operatorname{ord}_{\mathfrak{p}}g(P)\cdot\log N\mathfrak{p},$$

where g is a function which locally defines the divisor D, and $\mathfrak{p}$ is the prime ideal corresponding to the valuation v. Thus the points inside $\mathbb{D}_r$, correspond to non-archimedean places, and the summands (for fixed $w\in\mathbb{D}_r$ or fixed v) take on discrete sets of values.

Finally, we again let

$$\begin{aligned}T_D(P)&=m(D,P)+N(D,P)\\&=\frac{1}{[k:\mathbb{Q}]}\sum_{v}-\log\operatorname{dist}_v(D,P)\\&=h_D(P),\end{aligned}$$

which is a well-known definition in number theory known as the Weil height.

As before, we can define the defect $\delta(D)=\liminf m(D,P)/h_D(P)$. The assumption that D is defined over k implies that $\delta(D)<1$.

Then the following theorem holds with either set of definitions above, replacing "?" by r or P, as appropriate.

Theorem (Second Main Theorem). *Let D be a reduced effective divisor on a curve C. Let A be an ample divisor on C, let K be a canonical divisor on C, and let $\epsilon > 0$ be given. Then for almost all "?",*

$$m(D, ?) + T_K(?) \leq \epsilon T_A(?) + O(1).$$

Of course, in the Nevanlinna case, this is true with $(1 + \epsilon) \log T_A(r)$ in place of $\epsilon T_A(r)$, but this is only conjectured in the number field case.

In the number field case, when $g = 0$ this is Roth's theorem, which is the following.

Theorem (Roth, 1955). *Let k be a number field; for each archimedean place v of k let $\alpha_v \in \bar{\mathbb{Q}}$ be given. Also let $\epsilon > 0$. Then for all but finiteiy many $x \in k$,*

$$\prod_{v|\infty} \min(1, |x - \alpha_v|_v) > \frac{1}{H(x)^{2+\epsilon}}.$$

Here $H(x) = \prod_v \max(1, |x|_v)$, so that $h_{\mathcal{O}(1)}(x) = (1/[k : \mathbb{Q}]) \log H(x)$.

To see how this theorem follows from the Second Main Theorem, let A be a divisor corresponding to $\mathcal{O}(1)$, let D be the union of all conjugates over k of all α_v, and take $-\log$ of both sides. For details, see ([**V 3**], 3.2).

When $g(C) > 1$, the Second Main Theorem is equivalent to Mordell's conjecture. Indeed, take $D = 0$, so that $m(D, P) = 0$, and we can take $A = K$ since K is ample. Then let $\epsilon < 1$; this gives a bound for $h_K(P)$, which is unbounded for infinite sets of rational points. This gives a contradiction. Conversely, if there are only finitely many rational points, then any statement will hold up to $O(1)$.

If the genus of C equals 1, then the Second Main Theorem corresponds to an approximation statement on elliptic curves proved by Lang.

Note that in number theory the Second Main Theorem is viewed as an upper bound on $m(D, P)$ instead of a lower bound on $N(D, P)$ as is the case in value distribution theory.

The fact that the Second Main Theorem of Nevanlinna theory has just one proof valid for all values of $g(C)$ suggests that the same should hold for number fields. This led to a new proof of the Mordell conjecture ([**V 4**] and [**V 5**]), using methods closer to Roth's. Work on obtaining a truly combined proof is progressing. This new proof led Faltings [**F**] to generalize the methods to give two new theorems:

Theorem (Faltings). *Let X be an affine variety, defined over a number fieid k, whose projective closure is an abelian variety. Then the set of integral points on X (relative to the ring of integers in k) is finite.*

Theorem (Faltings). *Let X be a closed subvariety of an abelian variety A. Assume that both are defined over k, and that X does not contain any translates of any nontrivial abelian subvarieties of A. Then the set $X(k)$ of k-rational points on X is finite.*

This is still an incomplete answer, because if X does contain a nontrivial translated abelian subvariety of A, then this theorem provides no information. Instead, the following conjecture should hold:

Conjecture (Lang, [**L 1**]). *Let X be a closed subvariety of an abelian variety A. Then $X(k)$ is contained in the union of finitely many translated abelian subvarieties of A contained in X.*

Before discussing this further, let us recall some facts about the geometry of this situation. For all that follows, assume that X is a closed subvariety of an abelian variety A.

Theorem (Ueno, ([**Ii**], Ch. 10, Thm. 10.13)). *There exists an abelian subvariety B of A such that the map $\pi : A \to A/B$ has the properties that $X = \pi^{-1}(\pi(X))$ and $\pi(X)$ is a variety of general type.*

The map $\pi|X$ is calied the **Ueno fibration**. It is called **trivial** if B is a point.

Theorem (Kawamata Structure Theorem, [**K**]). *There exists a finite set $Z_1, \ldots, Z_n$ of subvarieties of X, each having nontrivial Ueno fibration, such that any nontrivial translated abelian subvariety of A contained in X is contained in one of the Z_i.*

The set $Z_1 \cup \cdots \cup Z_n$ is called the **Kawamata locus** of X.

Then Lang's conjecture is the analogue of the following statement, proved by Kawamata [**K**], using work of Ochiai [**O**]:

Theorem. *Let $f : \mathbb{C} \to X$ be a nontrivial holomorphic curve. Then the image of f is contained in the Kawamata locus of X.*

By the Kawamata Structure Theorem, this statement is equivalent to Bloch's conjecture, which asserts that the image of f is not Zariski-dense in X unless X itself is a translated abelian subvariety of A. (Bloch's conjecture was also proved, independently, by Green and Griffiths [**G-G**], also using Ochiai's work). Similarly, to prove Lang's conjecture, it would suffice to prove that $X(k)$ is not Zariski-dense unless X is a translated abelian subvariety of A.

For further details on these ideas, see ([**L 3**], Ch. 1 §6 and Ch. 8 §1). For more details on the connection with diophantine questions, see [**V 3**], especially Section 5.ABC for connections with the asymptotic Fermat conjecture.

Bibliography

[**F**] G. Faltings, Diophantine approximation on abelian varieties. *Ann. Math.*, **133** (1991) 549–576.

[**G-G**] M. Green and P. Griffiths, Two applications of algebraic geometry to entire holomorphic mappings. *The Chern Symposium 1979 (Proceedings of the International Symposium on Differential Geometry in Honor of S.-S. Chern, held in Berkeley, California, June 1979)*, Springer-Verlag, New York, 1980, 41–74.

[**Ii**] S. Iitaka, *Algebraic geometry: an introduction to the birational geometry of algebraic varieties* (Graduate texts in mathematics 76) Springer-Verlag, New York-Heidelberg-Berlin, 1982.

[**K**] Y. Kawamata, On Bloch's conjecture. *Invent. Math.*, **57** (1980), 97–100.

[**L 1**] S. Lang, Integral points on curves. *Publ. Math. IHES*, **6** (1960), 27–43.

[**L 2**] ———, *Fundamentals of diophantine geometry*, Springer-Verlag, New York, 1983.

[**L 3**] ———, *Diophantine geometry.* Encyclopedia of Mathematics, Springer-Verlag, New York, 1991.

[**O**] T. Ochiai, On holomorphic curves in algebraic varieties with ample irregularity, *Invent. Math.*, 43 (1977), 83–96.

[**V 1**] P. Vojta, A higher dimensional Mordell conjecture, in *Arithmetic Geometry*, ed. by G. Cornell and J. H. Silverman, Graduate Texts in Mathematics, Springer-Verlag, New York, 1986, 341–353.

[**V 2**] ———, A diophantine conjecture over $\overline{\mathbb{Q}}$, in *Séminaire de Théorie des Nombres, Paris 1984–85*, ed. by Catherine Goldstein, Progress in Mathematics 63, Birkhäuser, Boston-Basel-Stuttgart, 1986, 241–250.

[**V 3**] ———, *Diophantine approximations and value distribution theory*. (Lect. notes math., vol. 1239), Springer-Verlag, Berlin-Heidelberg-New York, 1987.

[**V 4**] ———, Mordell's conjecture over function fields. *Invent. Math.*, **98** (1989) 115–138.

[**V 5**] ———, Siegel's theorem in the compact case, *Ann. Math.*, **133** (1991) 509–548.

[**V 6**] ———, Arithmetic and hyperbolic geometry, in: *Proceedings of the International Congress of Mathematicians*, 1990, Kyoto, Japan, to appear.

DIOPHANTINE APPROXIMATION AND THE THEORY OF HOLOMORPHIC CURVES

Pit-Mann Wong

In the last few years, due to the works of Osgood [**O1,2**], Lang [**L1,2,3**], Vojta [**V1,2,3,4**] and others, there appear to be evidences that the Theory of Diophantine Approximation and The Theory of Holomorphic curves (Nevanlinna Theory) may be somehow related. Currently, the relationship between the two theories is still on a formal level even though the resemblance of many of the corresponding results is quite striking. Vojta has come up with a dictionary for translating results from one theory into the other. Again the dictionary is essentially formal in nature and seems somewhat artificial at this point, it is perhaps worthwhile to begin a systematic investigation. Recently, I began to study the Theory of Diophantine Approximations, with the motivation of formulating the theory so that it parallels the theory of curves. These notes is a (very) partial survey of some of the results in diophantine approximations and the corresponding results in Nevanlinna Theory.

(I) Diophantine Approximation

The theory of diophantine equations is the study of solutions of polynomials over number fields. Typically, results in diophantine equations come in the form of certain finiteness statements; for instance statements asserting that certain equations have only a finite number of rational or integral solutions. We begin with a simple example.

Example 1 Consider the algebraic variety $X^2 + Y^2 = 3Z^2$ in $\boldsymbol{P}^2$, we claim that there is no rational (integral) points (points with rational (integral) coordinates; on projective spaces a rational point is also an integral point) on this variety. To see this, suppose $P = [x, y, z]$ be a rational point on the variety with $x, y, z \in \boldsymbol{Z}$ and $gcd(x, y, z) = 1$.

This research was supported in part by the National Science Foundation Grant DMS-87-02144.

Then $x^2 + y^2 = 0$ (mod 3) so that $x = y = 0$ (mod 3). Thus x^2 and y^2 are divisible by 9 and it follows that z is divisible by 3, contradicting the assumption that $gcd(x, y, z) = 1$. This example illustrates one of the fundamental tools in diophantine equation:

> *"To show that a variety has no rational point, it is sufficient to show that the homogenous defining equation has no non-zero solutions mod p for one prime p"*

The converse to this statement, the so called "*Hasse Principle*" is not valid in general. The following example, due to Selmer:

$$3X^2 + 4Y^2 + 5Z^3 = 0$$

has no rational points and yet for any prime p, the corresponding equation mod p admits non-trivial solutions.

Example 2 The algebraic set $y^2 = x^3 + 17$ in A^2 has many rational points, for example (-2, 3), (-1, 4), (2, 5), (4, 9), (8, 23), (43, 282), (52, 375), (5234, 378661) are integral points; (-8/9, 109/27); (137/64, 2651/512) are rational points (unlike the projective varieties, a rational point on an affine variety may not be integral). In fact $V(\boldsymbol{Q})$ is infinite. If we homogenize the equation (replace x by X/Z, y by Y/Z), we get

$$Y^2Z = X^3 + 17Z^3$$

this defines a variety in $\boldsymbol{P}^2$. It has one point at infinity: $[0, 1, 0]$. The rational points are given by $\{(x, y) \in \boldsymbol{A}^2(\boldsymbol{Q}) | y^2 = x^3 + 17\} \cup \{[0, 1, 0]\}$. It can be shown that the line connecting any two $\boldsymbol{Q}$-rational points intersects the variety again in a $\boldsymbol{Q}$-rational point. In this way one can show that there are infinitely many $\boldsymbol{Q}$-rational points. The variety is an example of an elliptic curve. There are two fundamental theorems concerning elliptic curves: The Mordell-Weil theorem asserts that the set of rational points on an elliptic curve is finitely generated. The Siegel theorem asserts that the set of integral points on an elliptic curve is finite. In this example there are exactly 16 integral points, consisting of the eight points listed above and their negative (negative in the sense of the group law of an elliptic curve). For further discussions concerning elliptic curves we refer the readers to [**Sil**].

Example 3 Consider the equation

$$x^3 - 2y^3 = n$$

where n is any fixed integer. We claim that such an equation has only a finite number of integral solutions. First we reduce the problem to an estimate.

The left hand side of the equation can be factorized as

$$(x - \sqrt[3]{2}y)(x - \theta\sqrt[3]{2}y)(x - \theta^2\sqrt[3]{2}y)$$

where θ is the primitive cube root of unity. Dividing the equation by y^3 we get

$$\left(\frac{x}{y} - \sqrt[3]{2}\right)\left(\frac{x}{y} - \theta\sqrt[3]{2}\right)\left(\frac{x}{y} - \theta^2\sqrt[3]{2}\right) = \frac{n}{y^3}.$$

Since θ is non-real, the absolute value of the second and third terms on the left above are clearly bounded away from zero and we can choose the lower bound to be independent of x and y (take the smaller of the distances to the real axis from $\theta\sqrt[3]{2}$ and $\theta^2\sqrt[3]{2}$ for instance), so that

$$\left|\frac{x}{y} - \sqrt[3]{2}\right| \leq \frac{C}{|y|^3} \tag{1}$$

for some constant C independent of x and y. The problem is reduced to the problem of approximating irrational numbers by rationals. We shall see shortly that the inequality above can have only finitely many rational solutions.

First we recall a classical result of Liouville (cf. [Schm 2]).

Theorem 1 (Liouville 1851) *Let α be an algebraic number of degree $d \geq 2$ over $\boldsymbol{Q}$; i.e., $[\boldsymbol{Q}(\alpha) : \boldsymbol{Q}] = d$. Then there is a constant $C > 0$ (depending on α) so that for any rational number p/q (p, q integers and $q > 0$),*

$$\left|\frac{p}{q} - \alpha\right| \geq \frac{C}{q^d}. \tag{2}$$

Proof. Let $f(X) \in Z[X]$ be the minimal polynomial (of degree d) of α. For any rational number p/q, clearly $q^d f(p/q)$ is a non-zero integer (non-zero because f has no rational roots). Thus we get a lower bound for $|f(p/q)|$:

$$|f(p/q)| \geq 1/q^d \tag{3}$$

On the other hand, if $|p/q - \alpha| \leq 1$ then we can estimate $|f(p/q)|$ from above:

$$|f(p/q)| = |f(p/q) - f(\alpha)| = |f'(c)|\, |p/q - \alpha| \leq C'|p/q - \alpha|$$

where $C' = \sup_{|x-\alpha|\leq 1} |f'(x)|$. Combining with (3) we get

$$|p/q - \alpha| > C/q^d$$

where $C = 1/C'$, as claimed. If $p/q - \alpha| > 1$ then the theorem is trivially verified by taking $c = 1$ for instance. QED

Remark 1 The assumption that α be algebraic is crucial. In fact, this theorem is used by Liouville to construct transcendental numbers. For example, let

$$\alpha = \sum\nolimits_{1\leq n<\infty} 2^{-n!}, p_k = 2^{k!} \sum\nolimits_{1\leq n\leq k} 2^{-n!}, q_k = 2^{k!}$$

then

$$\left|\frac{p_k}{q_k} - \alpha\right| = \sum\nolimits_{k+1\leq n<\infty} 2^{-n!} < 2q_k^{-k-1} < cq_k^{-d}$$

for any given c and d, for all k sufficiently large. Thus α cannot be algebraic by the theorem. For more details concerning *Liouville numbers* and criterion of transcendence see Mahler [**Ma**] and Gelfond [**Ge**].

Remark 2 Liouville's theorem implies the following statement. Let α be an algebraic number of degree $d \geq 2$, then for any $\epsilon > 0$ there are at most finitely many rational numbers p/q (p, q integers and $q > 0$) such that

$$\left|\frac{p}{q} - \alpha\right| \leq \frac{1}{q^{d+\varepsilon}} \tag{4}$$

Suppose otherwise, then there are rational numbers p/q with arbitrary large q satisfying (4). Such rational numbers clearly violates (2) because for any $c > 0$, $q^{-(d+\varepsilon)} < cq^{-d}$ for q sufficiently large.

Returning to the example, the number $\alpha = \sqrt[2]{3}$ is algebraic of degree $d = 3$. Comparing inequality (1) with (2) we see that Liouville's theorem is almost but not quite strong enough to guarantee the finiteness of integral solutions of the equation in example 3. Before we recount the history of the improvements of Liouville's theorem, we should mention that for algebraic numbers of degree $d = 2$, Liouville's estimate is essentially sharp. This is a consequence of a very well-known result of Dirichlet ([**Schm2**]):

Theorem 2 (Dirichlet 1842) *For any irrational number α there exists infinitely many rational numbers p/q (p, q integers and $q > 0$) such that*

$$\left|\frac{p}{q} - \alpha\right| \leq \frac{1}{q^2}.$$

Remark 3 It follows that there are infinitely many rationals p/q with p and q relatively prime and satisfy the estimate above.

Remark 4 Dirichlet's theorem holds for any irrational number, algebraic or transcendental.

Thus for algebraic number α of degree 2, the exponent in (4) cannot be improved to 2. For algebraic numbers of higher degree the exponent $d + \varepsilon$ was improved to

$$\begin{array}{ll} 1 + \frac{1}{2}d & \text{(Thue, 1901)} \\ 2\sqrt{d} + \varepsilon & \text{(Siegel, 1921)} \\ \sqrt{2d} + \varepsilon & \text{(Dyson, also Gelfond, 1947)} \end{array}$$

and finally to $2 + \varepsilon$ by Roth (1955). Roth was awarded the Fields medal for this achievement.

Theorem 3 (Roth 1955) *Let α be an algebraic number of degree $d \geq 2$. Then for any $\varepsilon > 0$ the inequality*

$$\left|\alpha - \frac{p}{q}\right| \geq \frac{1}{q^{2+\varepsilon}}$$

holds with the exception of finitely many rationals p/q where p, q are integers and $q > 0$.

Remark 5 In the case where the degree of α is 2, Liouville's theorem is stronger than Roth's theorem.

Lang (**L1**]) conjectured that perhaps the estimate in Theorem 3 can be improved to

$$\left|\frac{p}{q} - \alpha\right| \geq \frac{1}{q^2 \log^{1+\varepsilon} q}.$$

The conjecture is still open at this time (the corresponding statement of this conjecture in Nevanlinna Theory is due to Wong [**2**], see also [**S-W**]). However, this estimate had been verified for some special numbers. There is also the theorem of Khinchin that this estimate holds for all but a set of numbers of zero Lebesgue measure (cf. Khinchin [**Kh**]). More precisely:

Theorem 4 (Khinchin) *Let φ be a positive continuous function on the positive real line such that $x\varphi(x)$ is non-increasing. Then for almost all (i.e., except on a set of zero Lebesgue measure) irrational number α, the inequality*

$$\left|\frac{p}{q} - \alpha\right| \geq \frac{\varphi(q)}{q}$$

holds for all but a finite number of solutions in integers p, q $(q > 0)$ if and only if the integral

$$\int_c^\infty \varphi(x)dx$$

converges for some positive constant c.

Using his improvement of Thue's theorem, Siegel proved the following finiteness theorem:

Theorem 5 (Siegel) *On an affine curve (over any number field) of positive genus there can only be a finite number of integral points.*

In the projective case, Mordell proved that

Theorem 6 (Mordell) *The set of rational points on an elliptic curve (i.e., a curve of genus one) is a finitely generated abelian group.*

Mordell also made the famous conjecture (solved by Falting in the affirmative, cf. *Endlichkeitssätze für abelsche Varietäten über Zahlkörpern*, Invent. Math. **73**, 183):

Mordell's Conjecture *There are only finitely many rational points on a curve of genus* > 1.

Remark 5 Unlike the affine case, there is no distinction between integral and rational points on a complete curve.

Falting's original proof of the Mordell's conjecture is geometric and did not use the theorem of Thue-Siegel-Roth. Vojta (1988) proved the Mordell conjecture over function fields and, more recently Falting gave a proof of the general case of the Mordel conjecture, using the Thue-Siegel-Roth's theorem. So far (essentially) all the known results in diophantine equations are consequences of the Thue-Siegel-Roth's theorem.

The extension of Roth's theorem to approximation of p-adic numbers by algebraic numbers, handling several valuations at the same time, is due to Ridout ([**Ri**]) and Mahler ([**Ma**]). First we recall the product formula of Artin-Whaples. Let $\boldsymbol{k}$ be a number field and v a valuation on $\boldsymbol{k}$. Denote by $\boldsymbol{k}_v$ the completion of $\boldsymbol{k}$ with respect to v and by $n_v = [\boldsymbol{k}_v : \boldsymbol{Q}_v]$ the local degree. Define an absolute value associated to an archimedean valuation v by

$$\|x\|_v = |x| \qquad \text{if } \boldsymbol{K}_v = \boldsymbol{R}$$
$$\|x\|_v = |x| \qquad \text{if } \boldsymbol{K}_v = \boldsymbol{C}$$

If v is non-archimedean then v is an extension of p-adic valuation on $\boldsymbol{Q}$ for some prime p, the absolute value is defined so that

$$\|x\|_v = |x|_p^{n_v}$$

if $x \in \boldsymbol{Q} - \{0\}$. With these conventions, there exists a complete set $M_{\boldsymbol{k}}$ of inequivalent valuations on $\boldsymbol{k}$ such that the product formula is satisfied with multiplicity one; i.e.,

(Artin-Whaples) $$\prod_{v \in M_{\boldsymbol{k}}} \|x\|_v = 1$$

for all $x \in \boldsymbol{k} - \{0\}$. Extend $\| \ \|_v$ to the algebraic closure $\bar{\boldsymbol{k}}_v$ of $\boldsymbol{k}_v$.

Let $\boldsymbol{k}$ be a field of characteristic zero, denote by $\boldsymbol{k}[X]$ and $\boldsymbol{k}(X)$ the polynomial ring and the rational function field over $\boldsymbol{k}$ respectively. Fix an *irreducible* polynomial $p(X)$ in $\boldsymbol{k}(X)$, define the order at p of a rational function $r(X)$ in $\boldsymbol{k}(X)$ to be α if $r = p^\alpha s/t$ where s and t are polynomials relatively prime to p. A p-adic valuation on $\boldsymbol{k}(X)$ is defined by

$$|r|_p = e^{-(\operatorname{ord} r)(\deg p)}.$$

For $r(X)$ in $\boldsymbol{k}(X)$, there exists polynomials f and g in $\boldsymbol{k}[X]$, where g is not the zero element, such that $r = f/g$. Define a valuation on $\boldsymbol{k}(X)$ by

$$|r|_\infty = |f/g|_\infty = e^{\deg f - \deg g}.$$

Denote by $\boldsymbol{k}(X)_\infty$ the completion of $\boldsymbol{k}(X)$ with respect to the valuation $| \ |_\infty$ and $\boldsymbol{k}(X)_p$ the completion of $\boldsymbol{k}(X)$ with respect to $| \ |_p$. The Artin-Whaple Product Formula is satisfied for $\boldsymbol{k}(X)$ (and also its finite algebraic extension).

We now give an analytic interpretation of the product formula. Consider the special case of $\boldsymbol{k} = \boldsymbol{C}$, the complex number field, there is also the field $\mathfrak{M}$ of meromorphic functions defined on a domain G in the Riemann sphere $\boldsymbol{CP}^1$. If $G = \boldsymbol{CP}^1$ then $\mathfrak{M} = \boldsymbol{C}(X) =$ the field of rational functions in one variable. Fix a point $z_0 \neq \infty$ in G and for a function $f \in \mathfrak{M}$, we may write

$$f(z) = (z - z_0)^{\operatorname{ord}_{z_0} f} g(z)$$

where g is meromorphic with $g(z_0) \neq 0$ or ∞. For a positive constant c, a valuation is defined by

$$|f|_{z_0} = c^{\operatorname{ord}_{z_0} f}.$$

Thus $\operatorname{ord}_{z_0} f > 0$ if z_0 is a zero of f and $\operatorname{ord}_{z_0} f < 0$ if z_0 is a pole. If $z_0 = \infty$, we may write

$$f(z) = z^{-\operatorname{ord}_\infty f} g(z)$$

where g is meromorphic with $g(z_0) \neq 0$ or ∞. A valuation is defined by setting

$$|f|_\infty = c^{\operatorname{ord}_\infty f}.$$

In the case of $G = \boldsymbol{CP}^1$, the valuation $|f|_{z_0}$ coincides with $|f|_p$ where $p(z) = z - z_0$ is an irreducible polynomial and $|f|_\infty = |f|_{p_\infty}$. In this case it is clear that

$$\prod\nolimits_{w \in \boldsymbol{CP}^1} |f|_w = c^{\#\text{ of zeros} - \#\text{ of poles}}$$

and the Product Formula for $\boldsymbol{C}(X)$ is equivalent to the following well-known theorem in complex analysis:

> *"For a rational function on* $\boldsymbol{CP}^1$ *the number of zeros and the number of poles are equal"*

It is understood that the numbers of zeros and poles are counted with multiplicities.

The generalization of the above statement to meromorphic functions is the Argument Principle:

Argument Principle *Let f be a function meromorphic on a domain G containing the closed disk $\bar{\Delta}_r$ or radius r. Assume that there are no zeros nor poles on the boundary $\partial\Delta_r$ of the disk, then*

$$n(0,r) - n(\infty,r) = \frac{1}{2\pi i} \int_{\partial\Delta_r} \frac{f'}{f} dz$$

where $n(0,r)$ and $n(\infty,r)$ are respectively the number of zeros and poles of f inside Δ_r.

If f is a rational function then there are only finitely many zeros and poles of f. We may choose sufficiently large r so that all zeros and poles of f, with the exception of the point at infinity, are inside the open disk Δ_r. Then

$$\mathrm{I}(f) = -\frac{1}{2\pi i}\int_{\partial\Delta_r}\frac{f'}{f}dz = \mathrm{ord}_\infty f$$

and

$$e^{I(f)} = |f|_\infty.$$

With this interpretation, it is clear that we recover the Product Formula. Another way of relating zeros and poles of meromorphic functions is through Jensen's Formula which will be discussed in the next section.

Roth's Theorem can be restated as follows:

Theorem 7 *Let* $\mathbf{k}$ *be a number field (a finite algebraic extension of* $\mathbf{Q}$*), and* $\{a_v \in \bar{\mathbf{Q}} \mid v \in S\}$ *where* $\bar{\mathbf{Q}}$ *is the algebraic closure of* $\mathbf{Q}$ *and* S *is a finite set of valuations on k containing all the archimedean valuations. Then for any positive real numbers c and* ε*, the inequality*

$$\prod_{v\in S}\min\{1, \|x - a_v\|_v\} \geq c\,H(x)^{-(2+\varepsilon)}$$

holds for all but finitely many x *in* $\mathbf{k}$*. Here* H *is the (multiplicative) height.*

The analogue in function fields of Liouville's theorem is due to Mahler ([**Ma1**]). He also showed that Liouville's theorem cannot be improved if the characteristic of the field of constant $\boldsymbol{k}$ is positive.

Theorem 8 (Mahler) *Let* $\alpha = \alpha(X)$ *be an element of* $\mathbf{k}(X)_\infty$ *algebraic, of degree* $d \geq 2$*, over* $\mathbf{k}(X)$*. Then there exists a constant* C *such that*

$$\left|\alpha - \frac{p}{q}\right| \geq \frac{C}{|q|^d}$$

for any polynomials p and q (q not the zero element) $\in \mathbf{k}[X]$. *If the characteristic of* $\mathbf{k}$ *is positive, the exponent d cannot be improved.*

The p-adic case is due to Uchiyama ([**U**]):

Theorem 9 (Uchiyama) *Let* $\alpha = \alpha(X)$ *be an element of* $\mathbf{k}(X)_p$ *algebraic, of degree* $d \geq 2$, *over* $\mathbf{k}(X)$. *Then there exists a constant C such that*

$$\left|\alpha - \frac{r}{s}\right| \geq \frac{C}{(\max\{|r|_p; |s|_p\})^d}$$

for any polynomials r and s (q not the zero element) $\in \mathbf{k}[X]$. *If the characteristic of* $\mathbf{k}$ *is positive, the exponent d cannot be improved.*

However, if the characteristic of $\mathbf{k}$ is *zero*, the function field analogue of Roth's theorem is valid.

Theorem 10 (Uchiyama) *Assume that char* $\mathbf{k} = 0$. *Then*

(i)Let $\alpha = \alpha(X)$ *be an element of* $\mathbf{k}(X)_\infty$ *algebraic, of degree* $d \geq 2$, *over* $\mathbf{k}(X)$. *Then for any* $\varepsilon > 0$, *there exists a constant C such that*

$$\left|\alpha - \frac{p}{q}\right| \geq \frac{C}{|q|^{2+\varepsilon}}$$

for all but a finite number of pairs of polynomials p and q (q not the zero element) $\in \mathbf{k}[X]$.

(ii)Let $\alpha = \alpha(X)$ *be an element of* $\mathbf{k}(X)_p$ *algebraic, of degree* $d \geq 2$, *over* $\mathbf{k}(X)$. *Then for any* $\varepsilon > 0$, *there exists a constant C such that*

$$\left|\alpha - \frac{r}{s}\right| \geq \frac{C}{(\max\{|r|_p; |s|_p\})^{2+\varepsilon}}$$

for all but a finite number of pairs of polynomials r and s (q not the zero element) $\in \mathbf{k}(X)$.

In the positive characteristic case, Armitage ([**Ar**]) found a condition for which Roth's theorem holds.

Theorem 11 (Armitage) *Assume that char* $\boldsymbol{k} > 0$. *Then the conclusions of Uchiyama's theorem hold for those algebraic* α *which does not lie in a cyclic extension of* $\boldsymbol{k}(X)$.

Remark Armitage actually proved the theorem for fields more general than function fields. But the conditions on these fields are somewhat technical to state at this point. One of these conditions is that the Artin-Whaples product formula holds. In the language of Nevanlinna theory, this means that the First Main Theorem holds).

Roth's theorem is extended to simultaneous approximations by W. M. Schmidt ([**Schm1,2**]) and later by Schlickewei ([**Schl**]) to the non-archimedean case. First we recall some terminologies. For a linear form L of $(n+1)$-variables with algebraic coefficients, (we shall also identified L with a hyperplane of $\boldsymbol{P}^n$), the Weil function $\lambda_{v;L}$ is defined by

$$\lambda_{v;L}(x) = \frac{1}{[\boldsymbol{k}:\boldsymbol{Q}]} \log \frac{(n+1)\|L\|_v \max_{0\leq j\leq n}\{\|x_j\|_v\}}{\|L(x)\|_v}.$$

where for a linear form $L(x) = \sum_{0\leq j\leq n} a_j x^j$, $\|L\|_v = \max_{0\leq i\leq n} \{\|a_i\|_v\}$. Hence $\lambda_{v,L}(x) \geq 0$.

Given a hyperplane L of $\boldsymbol{P}^n$ and a point $x \in \boldsymbol{P}^n(\boldsymbol{k})$ but $x \notin L$, the *proximity* and *counting* functions are defined by

$$m(x,L) = \sum_{v\in S} \lambda_{v,L}(x); \quad N(x,L) = \sum_{v\notin S} \lambda_{v,L}(x).$$

Note that both the proximity and the counting functions are ≥ 0.

By the definition of height we have the analogue of the First Main Theorem in Nevanlinna Theory ([**V1**], [**R-W**]:

Theorem 12 (First Main Theorem) *If* L *is a linear form and* $L(x) \neq 0$, *then*

$$m(x,L) + N(x,L) = h(x) + \frac{1}{[\boldsymbol{k}:\boldsymbol{Q}]} \log \prod_{v\in \boldsymbol{M_k}} \frac{(n+1)\|L\|_v}{\|L(x)\|_v}$$

where $h(x)$ *is the logarithmic (additive) height.*

The Theorem below (cf. Schmidt [**Schm1**] Theorem 2) is an analogue of the Second Main Theorem of Nevanlinna Theory for holomorphic curves. We shall use the same notation for a linear form and the hyperplane it defines. The following version of the subspace theorem is due to Schlickelwei [**Schl**] (see also Schmidt [**Schm1**], Theorem 3) and [**Schm2**]). The formulation below is due to Vojta ([**Vo1**] Theorem 2.2.4).

Theorem 13 (Subspace Theorem) *Let* $\{L_{v,i} \mid v \in S, 1 \le i \le n+1\}$ *be linear forms in* n*-variables with algebraic coefficients. Assume that for each fixed* $v \in S$ *(a finite set of valuations on* $\boldsymbol{k}$ *containing all archimedean valuations) the* $n+1$ *linear forms* $L_{v,1}, \ldots, L_{v,n+1}$ *are linearly independent. Then for any* $\varepsilon > 0$ *there exists a finite set* $\mathcal{J}$ *of hyperplanes of* $\boldsymbol{k}^{n+1}$ *such that the inequality*

$$\prod_{v \in S} \prod_{1 \le i \le n+1} \frac{1}{\|L_{v,i}(x)\|_v} \le \{\textit{size}\ (x)\}^{\varepsilon}$$

holds for all S*-integral points* $x = (x_0, \ldots, x_n) \in \mathcal{O}_{S^{n+1}} - \cup_{L \in \mathcal{J}} L$. *Here*

$$\textit{size}\ (x) = \max{}_{v \in S} \max{}_{0 \le j \le n} \{\|x_j\|_v.\}.$$

It is more convenient to formulate the Subspace Theorem projectively and express the estimate in terms of height rather than size.

Theorem 14 *Let* $\{L_{v,i} \mid v \in S, 1 \le i \le q\}$ *be linear forms of* $(n+1)$*-variables (or hyperplanes in* $\boldsymbol{P}^n$*) with algebraic coefficients. Assume that for each fixed* $v \in S$*, the hyperplanes* $L_{v,1}, \ldots, L_{v,q}$ *are in general position. Then for any* $\varepsilon > 0$ *there exists a finite set* $\mathcal{J}$ *of hyperplanes of* $\boldsymbol{P}^n(\boldsymbol{k})$ *such that the inequality*

$$\sum_{1 \le i \le q} \sum_{v \in S} \lambda_{v, L_{v,i}}(x) \le (n+1+\varepsilon)\, h(x)$$

holds for all points $x \in \boldsymbol{P}^n(\boldsymbol{k}) - \cup_{L \in \mathcal{J}} L$.

The Subspace Theorem of Schmidt can be extended to the case of hyperplanes in *sub-general position* by using the Nochka weight (cf. [**No**]). This extension is due to Ru and Wong [**R-W**]. First we recall the definition of sub-general position due to Chen (cf. [**Ch**]).

Definition Let V be a vector space over $\boldsymbol{F}$ (a field of characteristic 0) of dimension (over $\boldsymbol{F}$) $k+1$. Denote by V^* the dual of V. For $1 \leq k \leq n < q$, a collection of non-zero vectors $A = \{v_1, \ldots, v_q\}$ in V^* is said to be in *n-subgeneral position* iff the linear span of any (distinct) $n+1$ elements of A is V^*. If $n = k$ the concept coincides with the usual concept of *general position.*

Remarks (*i*) It is clear that $\{v_1, \ldots, v_q\}$ is in n-subgeneral position iff $\{\alpha_1 v_1, \ldots, \alpha_q v_q\}$ is in n-subgeneral position where each α_j is a unit of $\boldsymbol{F}$ (i.e., $\alpha_j \in \boldsymbol{F} - \{o\}$). Denote by $\boldsymbol{P}(V^*)$ the projective space of V^*. Then the elements of $\boldsymbol{P}(V^*)$ are identified as hyperplanes of the projective space $\boldsymbol{P}(V)$. A collection of hyperplanes $\{a_j \in \boldsymbol{P}(V^*) \mid 1 \leq j \leq q\}$ is said to be in *n-subgeneral position* iff $\{v_1, \ldots, v_q\}$ is in *n-subgeneral position* where $v_j \in V^*$ satisfies $\boldsymbol{P}(v_j) = a_j$. For $n = k$ this concept agrees with the usual concept of hyperplanes in general position.

(*ii*) If $A = \{v_1, \ldots, v_q\}$ is in n-subgeneral position then it is also in m-subgeneral position for all $m \geq n$ provided that $m < q$.

(*iii*) Let $(b_j \in \boldsymbol{P}(W^*) \mid 1 \leq j \leq q\}$ be hyperplanes in general position, where W is a vector space over $\boldsymbol{F}$ of dimension $n+1$. Let V be a vector subspace of W of dimension $k+1$; then $A = \{a_j = b_j \cap \boldsymbol{P}(V) \mid 1 \leq j \leq q\}$ is a set of hyperplanes in $\boldsymbol{P}(V)$, not necessarily in general position but is in n-subgeneral position.

Lemma (Nochka-Chen) *Let $A = \{v_1, \ldots, v_q\}$ be a set of vectors in V^* in n-subgeneral position. Then there exists a function, called the Nochka weight associated to $A, \omega : A \to \boldsymbol{R}$ and a constant θ with the following properties:*

(*i*) $$\frac{k+1}{2n-k+1} \leq \theta \leq \frac{k+1}{n+1},$$

(*ii*) $0 \leq \omega(a) \leq \theta$ *for* $a \in A$,

(*iii*) $\sum_{a\in A} \omega(a) = k+1+\theta(\#A - 2n + k + 1)$,

(*iv*) *for any subset B of A with $\#\ B \leq n+1$, $\sum_{a\in A} \omega(a) \leq d(B) =$ dimension of the linear space spanned by elements of B.*

The generalization of The Subspace Theorem to the case of hyperplanes in sub-general position takes the following form (cf. [**R-W**]):

Theorem 15 *Let* $\{L_{v,i} \mid v \in S, 1 \leq i \leq q\}$ *be linear forms of* $(k+1)$*-variables (or hyperplanes in* $\boldsymbol{P}^k$*) with algebraic coefficients. Assume that for each fixed* $v \in S$*, the linear forms* $L_{v,1}, \ldots, L_{v,q}$ *are in* n*-subgeneral position* $(1 \leq k \leq n < q)$ *with associated Nochka weights* $\omega_{v,1}, \ldots, \omega_{v,q}$. *Then for any* $\varepsilon > 0$ *there exists a finite set* $\mathcal{J}$ *of hyperplanes of* $\boldsymbol{P}^k(\boldsymbol{k})$ *such that the inequality*

$$\sum\nolimits_{1 \leq i \leq q} \sum\nolimits_{v \in S} \omega_{v,i} \lambda_{v,L_{v,i}}(x) \leq (k+1+\varepsilon)\, h(x)$$

holds for all points $x \in \boldsymbol{P}^k(\boldsymbol{k}) - \cup_{L \in \mathcal{J}} L$.

In terms of the proximity function, Theorem 15 takes the following forms:

Corollary *Let* $\{L_1, \ldots, L_q\}$ *be linear forms in* $(k+1)$*-variables with algebraic coefficients, in* n*-subgeneral position (* $1 \leq k \leq n$ *and* $q > 2n-k+1$*). Then for any* $\varepsilon > 0$ *there exists a finite set* $\mathcal{J}$ *of hyperplanes of* $\boldsymbol{P}^k(\boldsymbol{k})$ *such that*

$$\sum\nolimits_{1 \leq i \leq q} \omega_i m(x, L_i) \leq (k+1+\varepsilon)\, h(x)$$

holds for all points $x \in \boldsymbol{P}^k(\boldsymbol{k}) - \cup_{L \in \mathcal{J}} L$ *and where* ω_i *are the Nochka weights.*

Corollary *Let* $\{L_1, \ldots, L_q\}$ *be linear forms in* $(k+1)$*-variables with algebraic coefficients, in* n*-subgeneral position* $(1 \leq k \leq n)$. *Given any* $\varepsilon > 0$, *there exists a finite set* $\mathcal{J}$ *of hyperplanes of* $\boldsymbol{P}^k(\boldsymbol{k})$ *such that*

$$\sum\nolimits_{1 \leq i \leq q} m(x, L_i) \leq (2n-k+1+\varepsilon)\, h(x)$$

holds for all points $x \in \boldsymbol{P}^k(\boldsymbol{k}) - \cup_{L \in \mathcal{J}} L$.

Corollary *Let* $\{L_1, \ldots, L_q\}$ *be hyperplanes of* $\boldsymbol{P}^n(\boldsymbol{k})$*, in general position. Then for any* $\varepsilon > 0$ *and* $1 \leq k \leq n$*, the set of points of* $\boldsymbol{P}^n(\boldsymbol{K})$ *such that*

$$\sum\nolimits_{1 \leq i \leq q} m(x, L_i) \geq (2n-k+1+\varepsilon)\, h(x)$$

is contained a finite union of linear subspaces, $\cup_{L\in\mathcal{J}}L$, *of dimension* $k-1$. *In particular, the set of points of* $\boldsymbol{P}^n(\boldsymbol{k}) - \cup_{1\le i\le q}L_i$ *such that*

$$\sum\nolimits_{1\le i\le q} m(x, L_i) \ge (2n+\varepsilon)\, h(x)$$

is a finite set of points.

For a finite subset S of $M_{\boldsymbol{k}}$ of valuations containing the set S_∞ of all archimedean valuations of $\boldsymbol{k}$. Denote by $\mathcal{O}_S$ the ring of S-*integers* of $\boldsymbol{k}$, i.e., the set of $x \in \boldsymbol{k}$ such that

$$\|x\|_v \le 1$$

for all $v \notin S$. A point $x = (x_1, \ldots, x_n) \in \boldsymbol{k}^n$ is said to be a S-*integral point* if $x_i \in \mathcal{O}_S$ for all $1 \le i \le n$. Let D be a *very ample effective divisor* on a projective variety V and let $1 = x_0, x_1, \ldots, x_N$ be a basis of the vector space:

$$\mathfrak{l}(D) = \{f \mid f \textit{ is a rational function on the variety } V \textit{ such that } f = 0 \textit{ or } (f) + D \ge 0\}.$$

Then $P \to (x_1(P), \ldots, x_N(P))$ defines an embedding of $V(\boldsymbol{k}) - D$ into the affine space $\boldsymbol{k}^N$. A point P of $V(\boldsymbol{k}) - D$ is said to be a D-*integral point* if $x_i(P) \in \mathcal{O}_S$ for all $1 \le i \le N$.

The following theorem of Ru-Wong extends the classical theorem of Thue-Siegel that $\boldsymbol{P}^1 - \{3 \text{ distinct points}\}$ has finitely many integral points:

Theorem 16 *Let* $\boldsymbol{k}$ *be a number field and* $H_1, \ldots, H_q$ *be a finite set of hyperplanes of* $\boldsymbol{P}^n(\boldsymbol{k})$, *assumed to be in general position. Let* $D = \sum_{1\le j\le q} H_i$, *then for any integer* $1 \le k \le n$, *the set of* D-*integral points of* $\boldsymbol{P}^n(\boldsymbol{k}) - D$ *is contained in a finite union of linear subspaces of* $\boldsymbol{P}^n(\boldsymbol{k})$ *of dimension* $k-1$ *provided that* $q > 2n-k+1$. *In particular, the set of* D-*integral points of* $\boldsymbol{P}^n(\boldsymbol{k}) - \{2n+1$ *hyperplanes in general position*$\}$ *is finite.*

More generally, let V be a projective variety, D a very ample divisor on V and let $\{\varphi_0, \ldots, \varphi_N\}$ be a basis of $\mathfrak{l}(D)$, such that

$$\Phi = [\varphi_0, \ldots, \varphi_N] : V \to \boldsymbol{P}^N$$

is an embedding of V into $\boldsymbol{P}^N$ with $V - D$ embedded in $\boldsymbol{k}^N$. We identify V with its image $\Phi(V)$. As an immediate consequence of the main theorem we also have:

Corollary *Let V be a projective variety, D a very ample divisor on V. Let $D_1, \ldots, D_q$ be divisors in the linear system $|D|$ such that $E = D_1 + \ldots + D_q$ has, at worst, simple normal crossing singularities. If $q > 2N - k + 1$ where $N = \dim \mathfrak{l}(D) - 1$ and $1 \leq k \leq n$, then the set of E-integral points of $V - E$ is contained in the intersection of a finite number of linear subspaces, of dimension $k - 1$, of $\boldsymbol{P}^N$ with V. In particular, the set of E-integral points of $V - E$ is finite if $q \geq 2N + 1$.*

(II) Theory of Holomorphic Curves

The Theory of holomorphic curves is the study of holomorphic maps from the complex plane into complex manifolds. More generally, one studies holomorphic maps between complex manifolds with the case of curves being the most difficult. This is due to the fact that the image of a holomorphic curve is usually of high codimension. Typically results in the theory of maps assert that, under appropriate conditions, every holomorphic map in a complex manifold $\boldsymbol{M}$ degenerates. The types of degeneracy range from the weakest form: "the image does not contain an open set" to the strongest form: "the image consists of one point". In between we have degeneration at a certain dimension. Namely, the image is contained in a complex subvariety of dimension p with $0 \leq p < n = \dim_{\boldsymbol{C}} \boldsymbol{M}$.

Manifolds with the property that every holomorphic curve $f : \boldsymbol{C} \to \boldsymbol{M}$ is constant is said to be *Brody-hyperbolic* [**B**]. If $\boldsymbol{M}$ is compact, then the concept of Brody-hyperbolic is equivalent to the concept of Kobayashi-hyperbolic [**K1**]. The following differential geometric description of Kobayashi-hyperbolicity is due to Royden [**Roy**]. Given a non-zero tangent vector $\xi \in \boldsymbol{T}_x\boldsymbol{M}$, the *infinitesimal Kobayashi-Royden (pseudo) metric* is defined by

$$0 \leq \boldsymbol{k_M}(\xi) = \inf \frac{1}{r}$$

where the inf is taken over all positive real numbers r such that there exists a holomorphic map $f : \Delta_r \to \boldsymbol{M}$ such that $f(0) = x$ and $f'(0) = \xi$. Here Δ_r is the disk of radius r in C. Alternatively,

$$\boldsymbol{k_M}(\xi) = \sup |t| \geq 0$$

where the sup is taken over all $t \in \boldsymbol{C}^*$ such that there exists a holomorphic map $f : \Delta \to \boldsymbol{M}$ with $f(0)$ and $f'(0) = t\xi$. A complex manifold $\boldsymbol{M}$ is said to be *hyperbolic at a point* x if there exists an open neighborhood U of x and a hermitian metric $ds_U{}^2$ on TU such that $\boldsymbol{k_M} \geq ds_U$ on TU. A complex manifold $\boldsymbol{M}$ is said to be *hyperbolic* if it is hyperbolic at every point. The Kobayashi pseudo-distance associate to $\boldsymbol{k_M}$ is defined by

$$d(x, y) = inf \sqrt{\boldsymbol{k_M}(\gamma'(t))}$$

where the *inf* is taken over all piecewise smooth curves joining x and y. The condition that $\boldsymbol{M}$ is Kobayashi-hyperbolic is equivalent to the condition that the Kobayashi pseudo-distance is a distance; i.e., $d(x, y) > 0$ if $x \neq y$. With this distance function $\boldsymbol{M}$ is a metric space and $\boldsymbol{M}$ is said to be complete if $\boldsymbol{M}$ is a complete metric space.

The infinitesimal Kobayashi metric satisfies $\boldsymbol{k_M}(t\xi) = |t|\boldsymbol{k_M}(\xi)$, hence it is a *Finsler metric*. It has the nice property that every holomorphic map is metric decreasing. Namely, if $f : \boldsymbol{M} \to \boldsymbol{N}$ is holomorphic then $\boldsymbol{k_N}(f_*\xi) \leq \boldsymbol{k_M}(\xi)$. In particular, every biholomorphic self map of $\boldsymbol{M}$ is an *isometry* of the Kobayashi metric.

The infinitesimal Kobayashi metric does not have very good regularity in general. In this direction we have the fundamental result of Royden [**R1**] that the infinitesimal Kobayashi metric is always upper semi-continuous. If it is complete hyperbolic then the metric is continuous. It is well-known that the poly-discs are complete-hyperbolic but the infinitesimal Kobayashi metric is not differentiable.

As mentioned above, for compact manifolds Kobayashi-hyperbolic is equivalent to Brody-hyperbolic. In general, Kobayashi-hyperbolic implies Brody-hyperbolic but the converse may not be true if $\boldsymbol{M}$ is non-compact. The example below is very well-known.

Example (Eisenman and Taylor) Let $\boldsymbol{M}$ be the domain in $\boldsymbol{C}^2$ given by

$$\boldsymbol{M} = \left\{(z, w) \in \boldsymbol{C}^2 \mid |z| < 1, |zw| < 1 \text{ and } |w| < 1 \text{ if } z = 0\right\}$$

Then $\boldsymbol{M}$ contains no complex lines; for if $f : \boldsymbol{C} \to \boldsymbol{M}$ is holomorphic then $\pi_1 \circ f$ is bounded (where π_1 is the projection onto the first coordinate), hence constant; now if $\pi_1 \circ f$ is constant then $\pi_2 \circ f$ (where π_2 is the projection onto the second coordinate) is bounded and so must be constant as well. However $\boldsymbol{M}$ is not Kobayashi-hyperbolic, because the Kobayashi distance of any point of the form $(0, w)$ from the origin is zero. This is evident by considering the connecting paths: for any positive integer n, let $f_{j,n} : \Delta \to \boldsymbol{M}, j = 0, 1, 2$, be holomorphic maps defined by $f_{0,n}(z) = (z, 0), f_{1,n}(z) = (1/n, nz)$ and $f_{2,n}(z) = (1/n + z/2, w)$. Then $f_{0,n}(0) = (0, 0) = p_0, f_{0,n}(1/n) = (1/n, 0) = p_1, f_{1,n}(0) = (1/n, 0) = p_1, f_{1,n}(w/n) = (1/n, w) = p_2, f_{2,n}(0) = (1/n, w) = p_2, f_{2,n}(-2/n) = (0, w)$. The Kobayashi distances between the points $0, 1/n$ and $-2/n$ on the unit disc approaches zero as n approaches ∞.

Intuitively speaking, for non-compact manifolds, the points at "infinity" plays a very important role. In fact Green **[Gn4]** showed that

Theorem 17 (Green) *Let D be a union of (possibly singular) hypersurfaces $D_1, \ldots, D_m$ hypersurfaces in a complex manifold $\boldsymbol{M}$. Then $\boldsymbol{M} - D$ is Kobayashi-hyperbolic if*

(*i*) *There is no non-constant holomorphic curve $f : \boldsymbol{C} \to \boldsymbol{M} - D$;*
(*ii*) *There is no non-constant holomorphic curve*

$$f : \boldsymbol{C} \to D_{i_1} \cap \ldots \cap D_{i_k} - (D_{j_1} \cup \ldots \cup D_{j_l})$$

for all possible choices of distinct indices so that $\{i_1, \ldots, i_k\} \cup \{j_1, \ldots, j_l\} = \{1, \ldots, m\}$.

An important special case of this is the theorem:

Corollary (Green) *The complement of q hyperplanes in general position in $\boldsymbol{CP}^n$ is Kobayashi-hyperbolic for $q \geq 2n + 1$. The number $2n + 1$ is sharp.*

Let $\{P_j(z_0, \ldots, z_n) \mid 1 \leq j \leq k\}$ be a set of homogeneous polynomials with coefficients in an algebraic number field $\boldsymbol{k}$. Let V be the common zeros of $P_j (i \leq j \leq k)$ in $\boldsymbol{CP}^n$. We assume that V is irreducible and smooth. Lang conjectured that if V is hyperbolic then for any finite extension $\boldsymbol{K}$ of the field $\boldsymbol{k}$, the set of rational points $V(\boldsymbol{K})$ in V over $\boldsymbol{K}$ (i.e., $V(\boldsymbol{K}) = \{(z_0, \ldots, z_n) \mid$ if there exists i so that $z_i \neq 0$ and $z_j/z_i \in \boldsymbol{K}$ for all $j\}$ is finite. If V is a hyperbolic affine algebraic manifold in $\boldsymbol{k}^N$ defined by polynomials $\{Q_j(z_1, \ldots, z_N) = 0, 1 \leq j \leq k\}$ then the set of integral points of V over $\boldsymbol{K}$ is finite. This conjecture is verified for the case of curves of genus $g \geq 2$ (Falting); for $V = \boldsymbol{P}^n - \{2n+1$ hyperplanes in general position$\}$ (Ru-Wong) and for $V =$ complement of an ample divisor of an abelian variety (Faltings).

Green ([**Gn3**]) *also proved that if the hyperplanes are distinct but not in general position, one can still conclude that the complement is Brody-hyperbolic, namely it contains no non-trivial holomorphic curve from* **C**. *The corresponding statement in number theory:*

> *"$\boldsymbol{P}^n(\boldsymbol{k}) - \{2n+1$ distinct hyperplanes$\}$ contains only finitely many integral points"*

is still open. The proof of Ru-Wong for the case of hyperplanes in general position involves an extension of the Siegel-Roth-Schmidt type estimate for which the general position assumption is necessary.

Returning to the discussion of hyperbolic manifolds, the following Theorem of Brody ([**B**]) is very important in constructing examples.

Theorem 18 (Brody) *Small smooth deformations of a compact hyperbolic manifold are hyperbolic.*

Thus the set of compact hyperbolic manifolds is open. The following example of Brody and Green shows that it is not a closed set in general.

Example (Brody-Green) The hypersurface in $\boldsymbol{CP}^3$ defined by

$$V_\varepsilon = \left\{ z_0^d + z_1^d + z_2^d + z_3^d + (\varepsilon z_0 z_1)^{d/2} + (\varepsilon z_0 z_2)^{d/2} = 0 \right\}$$

is hyperbolic for any $\varepsilon \neq 0$ and where $d \geq 50$ is an even integer. For $\varepsilon = 0$, V_0 is a Fermat variety which is clearly not hyperbolic. Note that by the Lefschetz theorem, V_ε is simply connected. Since V_0 is non-singular (it is the Fermat surface of degree d) it follows that V_ε is non-singular for small ε. A Fermat surface of any degree admits complex lines, for instance take μ and η be any d-th roots of -1, then $z_0 = \mu z_1$ and $z_2 = \eta z_3$ is a complex line in the Fermat surface of degree d.

For the non-compact case, the problem of deformation of hyperbolic manifolds is much more complicated, additional assumptions are needed. The concept of "hyperbolic embeddedness" is needed, we shall not get into this here. The readers are encouraged to look into the very interesting paper of Zaidenberg [**Z**].

Classically, hyperbolicity is studied via the behavior of curvature. The most well-known Theorem is the Schwarz-Pick-Ahlfors lemma:

Theorem 19 *Let M be a complex hermitian manifold with holomorphic sectional curvature bounded above by a (strictly) negative constant. Then* **M** *is Kobayashi-hyperbolic.*

For a Riemann surface, Milnor [**Mi**] (see also Yang [**Ya**]) observed that the classical condition on the holomorphic curvature can be relaxed to the condition that the curvature satisfies $K(r) \leq -(1+\varepsilon)/(r^2 \log r)$ asymptotically where r is the geodesic distance from a point. The following higher dimensional analogue of Milnor's result is due to Greene and Wu ([**G-W**] p.113 Theorem G′). A point O of a Kähler manifold $\boldsymbol{M}$ is called a pole if the exponential map at O is a diffeomorphism of the tangent space at O onto $\boldsymbol{M}$. Let r be the geodesic distance from O. Let S_r be the geodesic sphere and X be the outward normal. Then $Z = X - \sqrt{-1}JX$ is called the (complex) *radial direction.* The *radial curvature* is defined to be the sectional curvature of the plane determined by the radial direction. With these terminologies we can now state the Theorem of Greene and Wu:

Theorem 20 *Let $\boldsymbol{M}$ be a complex Kähler manifold with a pole such that* (*i*)*the radial curvature is everywhere non-positive and $\leq -1/(r^2 \log r)$ asymptotically,* (*ii*)*the holomorphic sectional curvature $\leq -1/r^2$ asymptotically. Then $\boldsymbol{M}$ is Kobayashi-hyperbolic.*

The next Theorem ([**K-W**]) gives a criterion of hyperbolicity without requiring information on the precise rate of decay of the curvature. A complex manifold $\boldsymbol{M}$ of complex dimension n is said to be *strongly q-concave* if there exists a continuous function φ on $\boldsymbol{M}$ such that (*i*)for all real numbers c the set $\{z \in \boldsymbol{M} \mid \varphi(z) \leq c\}$ is compact, (*ii*)the Levi form $i\partial\bar{\partial}\varphi$ is semi-*negative* and has at least $n - q$ *negative* eigenvalues (in the sense of distributions) everywhere outside a compact set. Alternatively, $\boldsymbol{M}$ is strongly q-concave if there exists a continuous function φ on $\boldsymbol{M}$ such that (*i*)for all real numbers c the set $\{z \in \boldsymbol{M} \mid \varphi(z) \geq c\}$ is compact, (*ii*)the Levi form $i\partial\bar{\partial}\varphi$ is semi-*positive* and has at least $n - q$ *positive* eigenvalues (in the sense of distributions) everywhere outside a compact set.

Theorem 21 (Kreuzman-Wong) *Let $\boldsymbol{M}$ be a complete Kähler manifold of complex dimension m such that both the holomorphic sectional curvature and the Ricci curvature are (strictly) negative. Assume that $\boldsymbol{M}$ is strongly 0-concave and that the universal cover is Stein then $\boldsymbol{M}$ is Kobayashi hyperbolic.*

A complete simply connected Riemannian manifold $\boldsymbol{M}$ of non-positive Riemannian sectional curvature is said to be a *visibility manifold* if any two points at infinity (denoted $\boldsymbol{M}(\infty)$) can be joined by a unique geodesic in $\boldsymbol{M}$. A complete simply-connected Riemannian manifold with sectional curvature bounded above by a negative number (i.e., $\boldsymbol{K} \leq -b^2$) is a visibility manifold. More generally a complete simply-connected Riemannian manifold with strictly negative sectional curvature (i.e., $\boldsymbol{K} < 0$) and radial curvature $-\boldsymbol{K}(r)$, from some fixed point, satisfying the condition

$$\int_1^\infty r\boldsymbol{K}(r)\,dr = \infty$$

is a visibility manifold.

Corollary (Kreuzman-Wong) *Let $\boldsymbol{M}$ be a complete Kähler manifold such that its universal cover satisfies the visibility axioms and that the Riemannian sectional curvature satisfies $-a^2 \leq \boldsymbol{K} < 0$. Assume that $\boldsymbol{M}$ admits a finite volume quotient. Then $\boldsymbol{M}$ is Kobayashi hyperbolic.*

The proof of Theorem 21 (and the corollary) relies on the compacification theorem (again, we see that the "infinity" plays a crucial role) of Nadel and Tsuji [**N-T**] extending the result of Siu and Yau [**S-T**] on compacification of Kähler manifolds of finite volume and negative pinching (both above and below) of the Riemannian sectional curvature. The theorem of Siu and Yau gives more precise information on the compactification, in the case of Kähler manifolds, of the corresponding theorem of Gromov ([**B-G-S**]) in the Riemannian case. Both the theorem of Nadel-Tsuji and that of Siu-Yau have the origin in the work of Andreotti-Tomassini [**A-T**] on pseudoconcave manifolds. These theorems are natural generalization of the well-known compacification theorem for finite volume quotients of bounded symmetric domains.

Let D be an irreducible algebraic curve in $\boldsymbol{CP}^2$. At a point p in D let $A_1, \ldots, A_k$ be local irreducible components of D containing p. Let L be a projective line through p and denote by $m_j = \min_L$ $\{$intersection multiplicity of $L \cap A_j\}$. Then $(m_1 - 1, \ldots, m_k - 1)$ are the orders of irreducible singularities at p. Let

$$b = \sum_{1 \leq j \leq q} (m_j - 1)$$

be the total order of singularities of D. Denote by D^* the dual curve of D. The curve D is birationally equivalent to its dual D^*. The normalization of D and D^* are isomorphic. Denote by b^* the total order of singularities of D^*.

Theorem 22 (Grauert-Peternell [**G-P**]) *Let D be an irreducible algebraic curve in $\boldsymbol{CP}^2$ of genus $g \geq 2$. Assume that $b^* + \chi(D) < 0$ (where b^* is the total order of irreduible singularities of D) and that every tangent of D^* intersects D^* in at least two points. Then $\boldsymbol{CP}^2 - D$ is hyperbolic.*

At present one of the major open problems in the theory of hyperbolic manifolds is the following conjecture.

Conjecture : *For a generic algebraic curve D of degree $d \geq 5$ in $\boldsymbol{CP}^2$, the complement $\boldsymbol{CP}^2 - D$ is Kobayashi-hyperbolic.*

The space of algebraic curves of degree in d in $\mathbf{CP}^2$ is a projective manifold, denoted $\mathfrak{F}_d$. By a generic curve of degree d, we mean an element of $\mathfrak{F}_d - S_d$ where S_d is a subvariety of lower dimension. The conjecture is of course false without the "generic" condition. One of the difficulty of the conjecture is to describe the exceptional subvariety S_d. The interested readers are refer to the paper of Grauert [**G**] for many interesting ideas.

We now turn to the Second Main Theorem of Value Distribution Theory. It all begins from the fundamental work of Nevanlinna in one complex variable. Let $f : \mathbf{C} \to \mathbf{CP}^1$ be a holomorphic map. The characteristic function $T(f,r)$ is defined by

$$T(f,r) = \int_s^r \frac{dt}{t} \int_{\Delta_t} f^*\omega$$

where $0 < s \leq r$ and ω is the Fubini-Study metric on $\mathbf{CP}^1$. For a point $a \in \mathbf{CP}^1$, denote by $n(f,a,r)$ the number of preimages (counting multiplicities) of a inside the disk of radius r. The counting function $N(f,a,r)$ is defined by

$$N(f,a,r) = \int_s^r \frac{n(f,a,t)}{t} dt$$

The *proximity function* $m(f,a,r)$ is defined by

$$m(f,a,r) = \int_{\partial\Delta_r} \log \|f;a\| \frac{d\theta}{2\pi}$$

where $\|x;a\| = |<x,a>|/\|x\|\,\|a\|$ is the projective distance between x and a. Here $\|x\|^2 = |x_0|^2 + |x_1|^2$, $\|a\|^2 = |a_0|^2 + |a_1|^2$ and $<x,a> = x_0a_0 + x_1a_1$. The characteristic function, counting function and proximity function are related by the First Main Theorem of Nevanlinna ([**Ne1,2,3**]).

Theorem 23 (FMT) *Let $f : \mathbf{C} \to \mathbf{CP}^1$ be a non-constant holomorphic map and let a be a point of $\mathbf{CP}^1$. Then*

$$m(f,a,r)+N(f,a,r)=T(f,r)+O(1).$$

The FMT plays a similar role in Nevanlinna Theory as the role played by the Artin-Whaple product formula in Number Theory. The counterpart of Roth's Theorem in Number Theory is The Second Main Theorem of Nevanlinna:

Theorem 24 (SMT) *Let $f : \boldsymbol{C} \to \boldsymbol{CP}^1$ be a non-constant holomorphic map and $\{a_1, \ldots, a_q\}$ be a finite subset of $\boldsymbol{CP}^1$. Then for any $\varepsilon > 0$ there exists a set A of finite Lebesgue measure such that the following estimate holds for all $r \in [s, \infty) - A$:*

$$\sum\nolimits_{1\le j\le q} m(f,a_j,r) \le (2+\varepsilon)T(f,r).$$

Note that, by the FMT, the left hand side of the SMT can be replaced by $qT(f,r) - \sum_{1\le j\le q} N(f,a_j,r)$.

For a point $a \in \boldsymbol{CP}^1$, the defect $\delta_f(a)$ is defined to be

$$\delta_f(a) = \liminf_{r\to\infty} \frac{m(f,a,r)}{T(f,r)} == 1 - \limsup_{r\to\infty} \frac{N(f,a,r)}{T(f,r)}.$$

Corollary *Let $f : \boldsymbol{C} \to \boldsymbol{CP}^1$ be a non-constant holomorphic map. Then for a finite set $\{a_1, \ldots, a_q\}$ of $\boldsymbol{CP}^1$, the sum of defects satisfies the estimate*

$$\sum\nolimits_{1\le j\le q} \delta_f(a_j) \le 2.$$

The factor $2+\varepsilon$ in the SMT correspondes to the exponent $2+\varepsilon$ in Roth's Theorem. The fact that $\boldsymbol{CP}^1$—{3 distinct point} is hyperbolic is a consequence of Nevanlinna's SMT. This corresponds to the fact in number theory that Thue-Siegel Theorem (the integral points of $\boldsymbol{P}^1(\boldsymbol{k})$—{3 distinct point} is finite) follows from Roth's Theorem. The proofs of these two statements are completely analogous (cf. Ru-Wong [**R-W**]).

Nevanlinna's Theorem can be extended from $\boldsymbol{CP}^1$ to arbitrary Riemann surface in the following form. Let $\boldsymbol{M}$ be a Riemann surface with hermitian metric

$$ds^2 = h\frac{\sqrt{-1}}{2\pi}dz \wedge d\bar{z}.$$

Denote by $\boldsymbol{R}$ the Gaussian curvature of h; i.e.,

$$\boldsymbol{R} = h^{-1}\frac{\partial^2}{\partial z \partial \bar{z}} \log h.$$

Theorem 25 *Let $\boldsymbol{M}$ be a Riemann surface with hermitian metric ds^2 and let $f : \boldsymbol{C} \to \boldsymbol{M}$ be a non-constant holomorphic map. Let $\{a_1, \ldots, a_q\}$ be a finite set of $\boldsymbol{M}$. Then for any $\varepsilon > 0$ there exists a set A of finite Lebesgue measure such that*

$$qT(f, ds^2, r) - \sum\nolimits_{1 \le j \le q} N(f, a_j, r) \le$$

$$2\int_s^r \frac{dt}{t} \int_{|z| \le t} (f^*\boldsymbol{R}) f^* ds^2 + \varepsilon T(f, ds^2, r)$$

holds for all $r \in [s, \infty) - A$. Consequently, the sum of defects satisfies the estimate

$$\sum\nolimits_{1 \le j \le q} \delta(f, a_j) \le 2 \limsup \int_s^r \frac{dt}{t} \int_{|z| \le t} (f^*\boldsymbol{R}) f^* ds^2 / T(f, ds^2, r).$$

Here the characteristic function $T(f, ds^2, r)$ is given by

$$T(f, ds^2, r) = \int_s^r \frac{dt}{t} \int_{|z| \le t} f^* ds^2.$$

If the Gaussian curvature $\boldsymbol{R}$ is constant $(= c)$ then

$$\int_s^r \frac{dt}{t} \int_{|z| \le t} (f^*\boldsymbol{R}) f^* ds^2 = cT(f, ds^2, r)$$

and Theorem 25 takes a simpler form:

Corollary *Same assumption as in Theorem 25 and assume that the Gaussian curvature* $\boldsymbol{R} = c$ *of the Riemann surface* $\boldsymbol{M}$ *is constant. Then for any* $\varepsilon > 0$ *there exists a set* A *of finite Lebesgue measure such that*

$$qT(f, ds^2, r) - \sum_{1 \le j \le q} N(f, a_j, r) \le (2c + \varepsilon) T(f, ds^2, r)$$

holds for all $r \in [s, \infty) - A$. *Consequently, the sum of defects satisfies the estimate*

$$\sum_{1 \le j \le q} \delta(f, a_j) \le 2\, c.$$

For the Riemann sphere with the Fubini-Study metric, the curvature $R = 1$ (parabolic); for the torus (elliptic), the canonical metric is a flat metric, i.e. $R = 0$; and for surface of genus ≥ 2 (hyperbolic) with the canonical metric the curvature $R = -1$. Thus

Corollary (*i*) *If* $\boldsymbol{M} = \boldsymbol{CP}^1$ *then* $\sum_{1 \le j \le q} \delta(f, a_j) \le 2$*;*
(*ii*) *If* $\boldsymbol{M} = \boldsymbol{T} =$ *torus then* $\sum_{1 \le j \le q} \delta(f, a_j) \le 0$, *in particular every non-constant holomorphic map from* $\boldsymbol{C}$ *into* $\boldsymbol{T}$ *is surjective;*
(*iii*) *If genus* $\boldsymbol{M} \ge 2$ *then there is no non-constant holomorphic map from* $\boldsymbol{C}$ *into* $\boldsymbol{M}$, *i.e.,* $\boldsymbol{M}$ *is hyperbolic.*

The corresponding Theorems in number theory assert that the following spaces contain only finitely many integral points over any number field $\boldsymbol{k}$:

(*i*) (Thue, Roth, Schmidt) $\boldsymbol{P}^1$ - $\{3$ distinct point$\}$;
(*ii*) (Siegel) $\boldsymbol{T}^1$ - $\{$one point$\}$;
(*iii*) (Mordell conjecture) compact Riemann surfaces of genus ≥ 1.

Nevanlinna's Theorem can also be generalized to higher dimension for holomorphic maps between equidimensional manifolds. This extension is due to Carlson-Griffiths [**C-G**] and Griffiths-King [**G-K**]. Let $f : \boldsymbol{C}^n \to \boldsymbol{M}^n$ be a holomorphic map into a projective manifold. Let D be an ample divisor on $\boldsymbol{M}$ represented as the zero set of a holomorphic section s of a holomorphic line bundle $\mathfrak{L}$ over $\boldsymbol{M}$. The proximity function is defined by

$$m(f, D, r) = \int\limits_{\partial B_r} \log \frac{1}{\|s(f(z))\|} d\sigma$$

where B_r is the ball of radius r in $\boldsymbol{C}^n$ and $d\sigma$ is the rotationally invariant measure of the boundary, normalized so that the volume of the boundary ∂B_r is 1. Specifically,

$$d\sigma = d^c \log \|z\|^2 \wedge (dd^c \log \|z\|^2)^{n-1}.$$

Let $\mathfrak{L}$ be a holomorphic line bundle over $\boldsymbol{M}$ and let h be a hermitian metric on $\mathfrak{L}$ with Chern form ρ. The characteristic function of f is defined by

$$T(f, \mathfrak{L}, r) = \int\limits_0^r \frac{dt}{t^{2n-1}} \int\limits_{B_r} f^* \rho \wedge \omega^{n-1}$$

where $\omega = dd^c \|z\|^2$.

We state the SMT of Carlson-Griffiths-King in the sharper form of Wong [**W2**] (see also Goldberg-Grinshtein [**G-G**], Lang [**L4**] and Cherry [**Ch**]):

Theorem 26 *Let $\mathfrak{L}$ be a positive holomorphic line bundle over a projective manifold of dimension n and $D_1, \ldots, D_q \in |\mathfrak{L}|$ be divisors such that $D = D_1 + \ldots + D_q$ is of simple normal crossing. Let $\mathfrak{K}^*$ be the dual of the canonical bundle of $\boldsymbol{M}$. Let $f : \boldsymbol{C}^n \to \boldsymbol{M}^n$ be a non-degenerate (Jacobian not identically zero or equivalently, the image contains a non-empty open set). Then for any $\varepsilon > 0$, there exists a set A of finite Lebesgue measure such that the estimate*

$$\begin{aligned}\sum\nolimits_{1\le j\le q} m(f,D_j,r) \le & T(f,\mathfrak{f}^*,r) + \log T(f,\mathfrak{l},r) \\ & + n(1+\varepsilon)\log\log T(f,\mathfrak{l},r) \\ & + \frac{1}{2}n(1+\varepsilon)\log\log\log T(f,\mathfrak{l},r) \\ & + \frac{1}{2}n(1+\varepsilon)\log\log r\end{aligned}$$

holds for all $r \in [s,\infty) - A$.

Corollary *With the same assumptions as in Theorem 25 and assume in addition that* f *is transcendental. Then*

$$\sum\nolimits_{1\le j\le q} \delta(f,D_j) \le c_1(\mathfrak{f}^*)/c_1(\mathfrak{l}).$$

Theorem 26 holds if one replaces $\boldsymbol{C}^n$ by an affine algebraic manifold $\boldsymbol{N}$ of dimension $m \ge n = \dim \boldsymbol{M}$ and under the same non-degeneracy assumption; namely, the image of the map f contains a non-empty open set. Stoll extended Theorem 26 to the case where the domain is a parabolic manifold. In this more general case, the right hand side of the estimate of Theorem 26 is more complicated; terms involving the Ricci curvature of the parabolic manifold also appears. We refer the readers to Stoll [**Sto2**] for details. *The corresponding statement in number theory of the estimate in Theorem 26 is conjectured by Lang. This sharper form of the Roth's Theorem is still open.*

Nevanlinna's Theory can also be extended to the non-equidimensional case under a much weaker non-degeneracy assumption. This case is much harder and much deeper; so far the only satisfactory result is the case of hyperplanes in $\boldsymbol{CP}^n$ even though there are some progress in the more general case. The main ideas of handling linearly non-degenerate holomorphic curves are contained already in Ahlfors [**A**] (also H. Weyl and J. Weyl [**W-W**]; for a different approach see Cartan [**Ca**]). Unlike the case of Nevanlinna and also the case of Carlson-Griffiths-King where the first derivative of the holomorphic map contains all the necessary informations needed; the linearly non-degenerate condition involves, for curves in $\boldsymbol{CP}^n$, derivatives of f of order up to n. The informations contained in the derivatives are related by the Plücker Formula. Ahlfor's Theory was extended by Stoll to the case of linearly non-degenerate meromorphic maps from $\boldsymbol{C}^m$ into $\boldsymbol{CP}^n$. Stoll realized that the associated maps in the higher dimensional case, unlike the case of curves, are in general only meromorphic rather than holomorphic. This is so even if the original map is assumed to be holomorphic. Thus it is necessary to develop the whole theory for meromorphic maps. The Theory of Ahlfors and Stoll was extended by Murray where the domain is assumed to be Stein, and Wong where the domain is assumed to be affine algebraic or parabolic. The following sharper form of the SMT is due to Stoll-Wong [**S-W**].

Theorem 26 *Let $\boldsymbol{M}$ be an affine algebraic manifolds of dimension m and $f : \boldsymbol{M} \to \boldsymbol{CP}^n$ be a linearly non-degenerate meromorphic map. Let $a_1, \ldots, a_q$ be hyperplanes of $\boldsymbol{CP}^n$ in general position. Then for any $\varepsilon > 0$ there exists a set A of finite Lebesgue measure such that*

$$\begin{aligned}\sum\nolimits_{1\le j\le q} m(f,a_j,r) \le &(n+1)T(f,r) + d_M\frac{n(n+1)}{2}\Big\{\log T(f,r)\\ &+(2+\varepsilon)\log\log T(f,r) + \frac{n-1}{2}\log^+ r\\ &+\frac{1}{2}(5+3\varepsilon)\log^+\log^+ r\Big\}\\ &+(\log^+\log^+\log^+ T(f,r)\\ &+O(\log^+\log^+\log^+ r)\end{aligned}$$

for all $r \in [s,\infty) - A$ and where d_M is the degree of M.

The SMT for linearly non-degenerate curves in $\boldsymbol{CP}^n$ correspondes to the subspace Theorem of Schmidt in number theory. The SMT can also be extended to the case of hyperplanes of $\boldsymbol{CP}^k$ in n-subgeneral position. The result is first conjectured by H. Cartan and is known as Cartan conjecture. The conjecture is first resolved in the affirmative by Nochka and also in the Thesis of Chen. The corresponding result was due to Ru and Wong using ideas from the works of Nochka and Chen.

Theorem 27 *Let **M** be an affine algebraic manifolds of dimension m and $f : \boldsymbol{M} \to \boldsymbol{CP}^k$ be a linearly non-degenerate meromorphic map. Let $a_1, \ldots, a_q$ be a hyperplanes of $\boldsymbol{CP}^k$ in n-subgeneral position ($k \leq n$). Then for any $\varepsilon > 0$ there exists a set A of finite Lebesgue measure such that*

$$\sum\nolimits_{1 \leq j \leq q} m(f, a_j, r) \leq (n + 1 + \varepsilon) T(f, r)$$

for all $r \in [s, \infty) - A$.

Actually it is possible to obtain a more precise estimate as in Theorem 26.

For holomorphic curves from $\boldsymbol{C}$ into Abelian varieties, a SMT was obtained by Ochiai [**Oc**] and also by Noguchi [**Nog**] using jet metrics. However these results are not in the best possible form. In a forth coming paper we shall treat the case of holomorphic curves in spaces of constant sectional curvature. A sharp form of the SMT can be obtained via the use of Plücker's formula and also the technique of Siu described below. Recently R. Kobayashi, using a rather different method seems to obtain a fairly sharp SMT in the case of holomorphic curves in Abelian varieties.

The main ingredients of the proof are: Green-Jensen's Formula, Nevanlinna's lemma and Plücker's Formula.

Green-Jensen Formula *Let φ be a function of class $\boldsymbol{C}^2$ or a plurisubharmonic function or a plurisuperharmonic function on $\boldsymbol{C}^n$. Then*

$$\int_s^r \frac{dt}{t^{2m-1}} \int_{\|z\| \leq t} dd^c[\varphi] \wedge (dd^c \|z\|^2)^{n-1} =$$

$$\frac{1}{2} \int_{\|z\|=r} \varphi d^c \log \|z\|^2 \wedge (dd^c \|z\|^2)^{n-1}$$

$$- \frac{1}{2} \int_{\|z\|=s} \varphi d^c \log \|z\|^2 \wedge (dd^c \log \|z\|^2)^{n-1}$$

where $dd^c[\varphi]$ denotes differentiation in the sense of distribution.

In particular the lemma applies to $\varphi = \log|f|$ where f is a meromorphic function.

To state the lemma of Nevanlinna we need a definition. A non-negative, non-decreasing function g defined on $[0, \infty)$ is called a *growth function* if for any $t_0 > 0$

$$\int_{t_0}^{\infty} \frac{1}{tg(t)} dt = c_0(g, t_0) < \infty.$$

A typical growth function is $g(t) = (\log(1+t))^{1+\varepsilon}$ where $\varepsilon > 0$.

Nevanlinna's Calculus Lemma *Let T be a non-negative, non-decreasing, absolutely continuous function defined on the interval $[s, \infty)$ where $s \geq 0$. Let g be a growth function. Then there exists a measurable subset $[s, \infty)$ with finite Lebesgue measure such that*

$$T'(r) \leq T(r)\, g(T(r))$$

holds for all $r \in [s, \infty) - A$.

This technical lemma is fundamental in all the estimates encountered in Nevanlinna theory. Another lemma which is of technical as well as theoretical importance is Plücker's Formula. Let S be a Riemann surface with hermitian metric h and $(\boldsymbol{M}, g)$ be a complex manifold of dimension n with constant sectional curvature c. Let $f : S \to \boldsymbol{M}$ be a holomorphic curve and f_k be the k-th associate curves. Assume that $f_k \not\equiv 0$ for $1 \leq k \leq n$. Define differential forms

$$\Theta_k = dd^c \log|\Lambda_k|^2 + (k+1)c\frac{\sqrt{-1}}{2\pi}g - \frac{k(k+1)}{2}\frac{\sqrt{-1}}{2\pi}h, \qquad 1 \leq k \leq n-1.$$

We may now state the Plücker Formula (cf. [**W3**]):

Plücker's Formula for Spaces of Constant Curvature *Let (M, g) be a hermitian manifold of constant curvature c and S a Riemann surface with hermitian metric h. Let $f : S \to \boldsymbol{M}$ be a holomorphic* *curve which is non-degenerate of order k; i.e., the associate curve* $f_k \not\equiv 0$. *Then*

$$\begin{cases} \operatorname{Ric}\Theta_1 = \Theta_2 - 2\Theta_1 + c\Omega_g \\ \operatorname{Ric}\Theta_k = \Theta_{k-1} + \Theta_{k+1} - 2\Theta_k;\ 2 \leq k \leq n-1 \end{cases}$$

on $S - \{\zeta \in S | \Lambda_k(\zeta) = 0\}$. *Note that* $\Theta_0 = \Theta_n \equiv 0$.

One of the main reasons that the case of curves in $\boldsymbol{CP}^n$ works well is that the associate (osculating) curves are holomorphic (to see this one can either follow the method of Ahlfors or Wong [**W3**]). If the metric connection of the target space is holomorphic then of course all higher derivatives of the curve are also holomorphic and á priori so are the associate curves (which are the wedge product of the derivatives). However, connections are usually not holomophic (almost never is, for details see Wong [**W3**]); for instance the connection of the Fubini-Study metric is not holomorphic. On the other hand, meromorphic connections do exist on projective varieties, hence osculating curves defined via these connections are also meromorphic. This is the main idea of Siu's SMT.

Theorem 28 *Let* $\boldsymbol{M}$ *be a projective surface (i.e., complex dimension 2) with a meromorphic connection D. Let* $\mathfrak{t}$ *be a holomorphic section of a holomorphic line bundle* $\mathfrak{F}$ *over* $\boldsymbol{M}$ *such that* $\mathfrak{t} \otimes D$ *is holomorphic. Let* $f : \boldsymbol{C} \to \boldsymbol{M}$ *be a holomorphic curve which is non-degenerate in the sense that the image of* f *is not contained entirely in the pole set of D and that* $f' \wedge Df' \not\equiv 0$. *Let* $\mathfrak{l}$ *be a holomorphic line bundle over* $\boldsymbol{M}$ *with a non-trivial holomorphic section s such that* $D = [s = 0]$ *is non-singular. Then for any* $\varepsilon > 0$ *and* $\lambda > 1$ *there exists a set A of finite Lebesgue measure such that*

$$(1 - \varepsilon)T(f, r) + N(f, D, r) \leq \lambda T(f, \mathfrak{f}^* \otimes \mathfrak{l}^*, r) + o(T(f, \mathfrak{l}, r)$$

for all $r \in [s, \infty) - A$.

Siu's Theorem provides some very interesting new examples even though this approach does not yet produce the "right" estimate in many of the important cases. The problem lies in the difficulty of controlling the pole order of the meromorphic connection, making optimal estimate in the SMT unattainable.

Another long standing problem which is solved only in the last few years is the problem of moving target. A hyperplane in $\boldsymbol{CP}^n = \boldsymbol{P}(\boldsymbol{C}^{n+1})$ may be identified with a point in the dual $\boldsymbol{P}((\boldsymbol{C}^{n+1})^*)$. But instead of considering fixed hyperplanes $a_1, \ldots, a_q \in \boldsymbol{P}((\boldsymbol{C}^{n+1})^*)$ one considers holomorphic curves $g_1, \ldots, g_q : \boldsymbol{C} \to \boldsymbol{P}((\boldsymbol{C}^{n+1})^*)$. In the one dimensional case, Nevanlinna conjectured that the deficit estimate of a holomorphic curve $f : \boldsymbol{C} \to \boldsymbol{CP}^1$ remains valid if the growth of characteristic functions $T(g_j, r)$ of the moving hyperplanes is slower than the growth of the characteristic function $T(f, r)$. Chuang [**Chu**] made significant progress on this problem. The conjecture is finally solved by Steinmetz in 1986 for curves into $\boldsymbol{CP}^1$. The case of curves in $\boldsymbol{CP}^n$ is solved by Ru-Stoll [**R-S1,2**] recently. Bardis [**Ba**] and O'Shea [**OS**] extended the Theory of moving targets to the case where the domain is also of higher dimension; deficit estimates are obtained under additional assumptions. We shall only state the SMT of Steinmetz and Ru-Stoll here. Given a family of holomorphic maps $\{g_1, \ldots, g_q\}$ from $\boldsymbol{C}$ into $\boldsymbol{P}((\boldsymbol{C}^{n+1})^*)$, *the field of meromorphic functions generated by* $\{g_1, \ldots, g_q\}$ is the smallest subfield $\mathfrak{B}$ of the field of meromorphic functions on $\boldsymbol{C}$ containing all the coordinate functions of $g_j, 1 \le j \le q$. A holomorphic curve $f : \boldsymbol{C} \to \boldsymbol{CP}^n$ is said to be linearly non-degenerate over $\mathfrak{B}$ if the coordinate functions of f does not satisfies any non-trivial linear equation with coefficients in $\mathfrak{B}$.

Theorem 29 *Let $f : \boldsymbol{C} \to \boldsymbol{CP}^n$ be a holomorphic curve and $g_1, \ldots, g_q : \boldsymbol{C} \to \boldsymbol{P}((\boldsymbol{C}^{n+1})^*)$ be q holomorphic maps considered as moving hyperplanes of $\boldsymbol{CP}^n$ in general position. Assume that $T(g_j, r)/T(f, r) \to 0$ as $r \to \infty$ and that f is linearly non-degenerate over $\mathfrak{B}$. Then for any $\varepsilon > 0$ there exists a set A of finite Lebesgue measure such that the estimate*

$$\sum_{1 \le j \le q} m(f, g_j, r) \le (n + 1 + \varepsilon) T(f, r)$$

holds for all $r \in [s, \infty) - A$. Consequently

$$\sum_{1 \le j \le q} \delta(f, g_j) \le n + 1.$$

In fact Ru-Stoll obtained a version of the SMT for moving targets corresponding to the Cartan conjecture. For this they introduce a concept called k-flat (we refer the readers to [**R-S2**] for details.

Theorem 30 *Let $f : \boldsymbol{C} \to \boldsymbol{CP}^n$ be a holomorphic curve and $g_1, \ldots, g_q : \boldsymbol{C} \to \boldsymbol{P}((\boldsymbol{C}^{n+1})^*)$ be q holomorphic maps considered as moving hyperplanes of $\boldsymbol{CP}^n$ in general position. Assume that $T(g_j, r)/T(f, r) \to 0$ as $r \to \infty$ and that the dimension of the map f is k-flat over $\mathfrak{B}$. Then for any $\varepsilon > 0$ there exists a set A of finite Lebesgue measure such that the estimate*

$$\sum_{1 \le j \le q} \omega_j m(f, g_j, r) \le (2n - k + 1 + \varepsilon)\, T(f, r)$$

holds for all $r \in [s, \infty) - A$ and where the ω_j's are the Nochka weights associated to the g_j's. Consequently

$$\sum_{1 \le j \le q} \delta(f, g_j) \le 2n - k + 1.$$

(III) Remarks

From the results listed in the two previous sections, the similarities between the two theories seem quite striking. The results in the Theory of curves are more complete due to the fact that there are more tools available. The analytic machineries are more powerful; the idea of Nevanlinna using invariants defined by integrals (e.g., characteristic functions, proximity functions) makes estimates easier to obtain (pointwise estimates are replaced by integral estimates). Furthermore, the proofs of the various Theorems in Nevanlinna Theory are quite uniform. The basic approach and the basic steps are essentially the same. The key ingredients are the Jensen formulas (corresponds to the Artin-Whaple's Product Formula), Ahlfors' Theory of associate (osculating) curves (corresponds to the successive minima in the geometry of numbers) and Nevanlinna's calculus lemma estimating the derivative of a positive convex increasing function by the function itself.

Even though Nevanlinna's lemma is elementary in nature, it has the effect of making many estimates routine. Without this technical lemma Nevanlinna Theory would be much more complicated. Unfortunately, there is as yet no good analog of Nevanlinna's lemma in Number Theory. This perhaps is the main reason that the proofs in diophantine approximation are not as uniform; many estimates are obtained via ingenious process which are perhaps not so "natural". The search of a good analog of Nevanlinna's lemma should be one of the main technical goal in the theory of diophantine approximations.

If one compares the theory of successive minima to the theory of associate curves one notices that the later is much more well-developed. The center-piece of the theory of associate curves is the Formula of Plücker, relating the invariants of higher order osculating curves to that of the lower order osculating curves. The counterpart of Plücker's Formula in the theory of successive minima has yet to be developed. The precise relations among the successive minima seem rather complicated at this point. A better understanding of these fundamental relationships would go a long way in developing the theory of diophantine approximations.

References

[**A**] Ahlfors, L. *The Theory of Meromorphic Curves,* Acta Soc. Sci. Fenn. Ser. A, vol. **3**, no. **4** (1941).

[**A-G**] Andreotti A. and H. Grauert, *Algebraische Körper von automorphen Funktionen,* Nachr. Akad. Wiss. Göttingen, Math-.-Phys. Kl. II (1961), 39–48.

[**Ar**] Armitage, *The Thue-Siegel-Roth Theorem in Characteristic p,* J. Alegbra **9**, 1968).

[**A-T**] A. Andreotti and G. Tomassini, *Some Remarks on Pseudoconcave Manifolds,* Essays on Topology and Related Topics (dedicated to G. de Rham), Springer-Verlag, 1970.

[**B**] R. Brody, *Compact manifolds and hyperbolicity,* Trans. Am. Math. Soc., **235** (1978), 213–219.

[**Ba**] Bardis, M., *Defect relation for meromorphic maps defined on covering parabolic manifolds,* Notre Name Thesis 1990.

[B-G-S] W. Ballman, M. Gromov and V. Schroeder, *Manifolds of Nonpositive Curvature,* Progress in Math. **61** (1985), Birkhäuser.

[Ca] Cartan, H., *Sur les zéros des combinaisons linéaires de p fonctions holomorphes données,* Mathematica (cluj) **7**, 80–103 (1933).

[C-G] Carlson, J. and Ph. Griffiths, *Defect relation for equidimensional holomorphic mappings between algebraic varieties,* Ann. of Math. **95** (1972), 557–584.

[Ch] Chen, W. *Cartan's Conjecture: Defect Relations for Meromorphic Maps from Parabolic Manifold to Projective Space,* University of Notre Dame Thesis 1987.

[Chu] Chuang, C.-T., *Uné généralization d'une inégalité de Nevanlinna*. Sci. Sinica **13** (1964), 887–895.

[Cy] Cherry, W., *The Nevanlinna error term for coverings generically surjective case,* preprint (1991).

[E-N] P. Eberlein and B. O'Neill, *Visibility Manifolds,* Pacific J. of Math. **46** (1973), 45–109.

[F1] Faltings, G. *Endlichkeitssätze für abelsche Varietäten über Zahlkörpern,* Invent. Math. **73**, 349–366 (1983).

[F2] ______, *Diophantine Approximations on Abelian Varieties,* Preprint (1989).

[G] Grauert, H. *Jetmetriken und hyperbolische Geometrie,* Math. Zeitschr. **200**, 149–168 (1989).

[Ge] Gelfond, *Transcendental and Algebraic Numbers,* Dover Publications, Inc. (1960).

[G-G] Goldberg, A. A. and V. A. Grinshtein, *The Logarithmic Derivative of a Meromorphic Function,* Math. Notes **19** 1976, AMS translation 320–323.

[G-K] Griffiths, Ph. and J. King, *Nevanlinna theory and holomorphic mappings between algebraic varieties,* Acta Math. **130** (1973), 145–220.

[G-U] Grauert, H. and U. Peternell, *Hyperbolicity of the complement of plane curves*. Manuscripta Math. **50** (1985), 429–441.

[Gn1] Green, M., *The complement of the dual of a plane curve and some new hyperbolic manifolds.* Value Distribution Theory,

Part A edited by Kujala and Vitter, Marcel Dekker, New York (1974), 119–131.

[GN2] ———, *Some examples and counter-examples in value distribution theory in several variables.* Compositio Math.

[Gn3] ———, *Some Picard theorems for holomorphic maps to alegbraic varieties.* Amer. J. Math. **97** (1975), 43–75.

[Gn4] ———, *Holomorphic maps into complex projective space omitting hyperplanes,* Trans. Am. Math. Soc. **169**, 89–103 (1972).

[G-W] Green R. and H. Wu, *Function Theory on Manifolds Which Possess a Pole,* Springer Lecture Notes **669** (1979).

[K1] Kobayashi, S. *Hyperbolic Manifolds and Holomorphic Mappings,* Marcel Dekker, New York, 1970.

[K2] ———, *Intrinsic distances, measures and geometric function theory,* Bull. Amer. Math. Soc., **82** (1976), 357–416.

[Kh] Khinchin, *Continued Fractions,* The University of Chicago Press, 1964.

[K-K1] Kiernan P. and S. Kobayashi, *Satake Compactification and Extension of Holomorphic Mappings,* Invent. Math. **16** (1972), 237–248.

[K-K2] ———, *Comments on Satake Compactification and the Great Picard Theorem,* J. Math. Soc. Japan **28** (1976), 577–580.

[K-W] Kreuzman, M. J. and P. M. Wong, *Hyperbolicity of negatively curved Kähler manifolds,* Math. Annalen. **287** (1990), 47–62.

[L1] Lang. S. *Report on diophantine approximations,* Bull. Soc. Math. France **93** (1965), 177–192.

[L2] ———, *Fundamentals of Diophantine Geometry,* Springer-Verlag, New York (1983).

[L3] ———, *Hyperbolic and Diophantine Analysis,* Bull. AMS, **14**, 159–205 (1986).

[L4] ———, *The error terms in Nevanlinna Theory,* Duke Math. J. **56**. (1988), 193–218.

[Ma1] Mahler, K. *On a Theorem of Liouville in Fields of Positive Characteristic,* Canadian J. Math. **1**, 1949.

[Ma2] ______, *Lectures on Diophantine Approximations,* University of Notre Dame Press (1961).

[M] Milnor, J., *On deciding whether a surface is parabolic or hyperbolic,* Amer. Math. Monthly, **84** (1977), 43–46.

[Ne1] Nevanlinna, R., *Einige Eindeutigkeitssatze in de Theorie der meromorphen Funktionen,* Acta Math. **48** (1926), 367–391.

[Ne2] ______, *Le Theoreme de Picard-Borel et la Theorie des Fonctions Meromorphes,* Chelesa Publ. Co. New York (1974).

[Ne3] ______, *Eindeutige analytische Functionen,* 2nd ed. Die Grundle. Math. Wiss. **46** (1953), Springer-Verlag.

[No] Nochka, E. I., *On the theory of meromorphic functions,* Soviet Math. Dokl,. **27** (2) (1983).

[N-T] Nadel, A. and H. Tsuji, *Compactification of complete Kähler manifolds of negative Ricci curvature,* J. Diff. Geom. **28** (1988), 503–512.

[O1] Osgood, C. F., *A number theoretic-differential equations approach to generalizing Nevanlinna theory,* Indiana J. of Math. **23**, 1–15 (1981).

[O2] ______, *Sometimes effective Thue-Siegel-Roth-Schmidt-Nevanlinna bounds, or better,* J. Number Theory **21**, 347–389 (1985).

[Oc] Ochiai, T., *On holomorphic curves in algebraic varieties with ample irregularity,* Invent. Math. **43**, 83–96 (1977).

[OS] O'Shea, A. Notre Dame Thesis, 1991.

[Ri] Ridout, The *p*-adic generalization of the Thue-Siegel-Roth theorem, Mathematika **5**, 1958.

[Ro] Roth, K. F., *Rational approximations to algebraic numbers,* Mathematika **2**, 1–20 (1955).

[Roy] Royden, H. *Remarks on the Kobayashi metric,* Proc. Maryland Conference on Several Complex Variables, Springer Lecture Notes **183** (1971).

[R-S1] Ru, M. and W. Stoll. *Courbes holomorphes évitant des hyperplans mobiles*. C. R. Acad. Sci. Paris t. **310** Série I. 45–48 (1990).

[R-S2] ______, *The Cartan Conjecture for Moving Targets*. Proc. Symp. Pure Math. **52** (1991) 477–507.

[**R-W**] Ru, M. and P.M. Wong, *Integral Points of* $\boldsymbol{P}^n(\boldsymbol{K}) - \{2n+1$ *hyperplanes in general position*$\}$, Invent. Math. **106** (1991), 195–216.

[**Sa**] Satake, I., *On compactifications of the quotient spaces for arithmetically defined discontinuous groups,* Ann. of Math. **72** (1960), 555–580.

[**Schl**] Schlickewei, H. P., *The p-adic Thue-Siegel-Roth-Schmidt theorem,* Archiv der Math. **29**, 267–270 (1977).

[**Schm1**] Schmidt, W. M., *Simultaneous Approximation to Algebraic numbers by Elements of a Number Field,* Monatshefte für Math. **79**, 55–66 (1975).

[**Schm2**] ———, *Diophantine Approximation,* Springer Lecture Notes **785**, Springer-Verlag, New York (1980).

[**Si**] Siegel, C. L., *Approximation algebraischer Zahlen,* Math. Zeitschr. **10**, 173–213 (1921).

[**Sil**] Silverman, J. H., *The Arithmetic of Elliptic Curves*, Grad. Texts in Math. **106**, Springer-Verlag (1986).

[**Siu**] Siu, Y.T., *Defect relation for holomorphic maps between spaces of different dimensions,* Duke Math. J. **55** (1987), 213–251.

[**Sh1**] Shiffman, B., *Introduction to Carlson-Griffiths equidistribution theory,* Springer Lecture Notes in Math. **981** (1983), 44–89.

[**Sh2**] ———, *A general second main theorem for meromorphic functions on* $\boldsymbol{C}^n$, Amer. J. of Math. **106** (1984), 509–531.

[**S-T**] Siu, Y.T. and S.-T. Yau, *Compactification of negatively curved complete Kähler manifolds of finite volume,* Seminar on Diff. Geom. edited by S. T. Yau, Ann. of Math Studies **102** (1982), 363–380.

[**Stm**] Steinmetz, N., *Eine Verallgemeinerung des zweiten Nevanlinnaschen Hauptsatzes,* J. Reine Angew. Math. **368** (1986), 134–141.

[**Sto1**] Stoll, W. *Die beiden Hauptsatze der Wertverteilungstheorie bei Funktionen mehrerer komplexen Verlanderlichen I,* Acta Math. **90** (1953), 1–115; *II,* Acta Math. **92** (1954), 55–169.

[**Sto2**] ———, *Value distribution on parabolic spaces,* Springer Lecture Notes in Math. **600** (1977).

[Sto3] ———, *Deficit and Bezout Estimates,* Valuation Distribution Theory, Part B, edited by R. O. Kujala and A. Vitter III, Marcel Dekker (1973).

[Sto4] ———, *Introduction to value distribution theory of meromorphic maps,* Springer Lecture Notes in Math. **950** (1982), 210–359.

[Sto5] ———, *The Ahlfors-Weyl theory of meromorphic maps on parabolic manifolds,* Springer Lecture Notes in Math. **981** (1983), 101–219.

[S-W] Stoll, W. and P. M. Wong, *Second Main Theorem of Nevanlinna Theory for Non-equidimensional Meromorphic Maps,* preprint 1991.

[T] Thue, A., *Über Annäherungswerte algebraischer Zahlen,* J. reine ang. Math. **135**, 284–305 (1909).

[U] Uchiyama, *Rational Approximations to Algebraic Functions,* J. Fac. Sci. Hokkaido Univ. **15**, 1961 p 173–192).

[V1] Vojta, P. *Diophantine Approximations and Value Distribution Theory,* Springer Lecture Notes **1239**, Springer-Verlag (1987).

[V2] ———, *Mordell's conjecture over function fields,* Invent. Math. **98**, 115–138 (1989).

[V3] ———, *Siegel's Theorem in the compact case,* preprint (1989).

[V4] ———, *A Refinement of Schmidt's Subspace Theorem,* Amer. J. Math. **111**, 489–518 (1989).

[Vi] Vitter, A., *The lemma of the logarithmic derivative in several complex variables,* Duke Math. J., **44** (1977), 89–104.

[W1] Wong, P. M., *Defect Relations for Maps on Parabolic Spaces and Kobayashi Metric on Projective Spaces Omitting Hyperplanes,* University of Notre Dame Thesis (1976).

[W2] ———, *On The Second Main Theorem of Nevanlinna Theory,* Amer. J. Math. **111**, 549–583 (1989).

[W3] ———, *On Holomorphic Curves in Spaces of Constant Holomorphic Sectional Curvature,* preprint 1990, to appear in Proc. Conf. in Complex Geometry, Osaka, Japan (1991).

[**W4**] ———, *Holomorphic mappings into Abelian Varieties,* Amer. J. Math. **102**, 493–501 (1980).

[**Wu**] Wu, H., *The equidistribution theory of holomorphic curves,* Ann. of Math Studies **64** (1970), Princeton University Press.

[**W-W**] Weyl, H. and J. Weyl, *Meromorphic functions and analytic curves,* Ann. of Math. Studies **#12**, Princeton University Press (1943).

[**Y**] P. Yang, *Curvature of complex submanifolds of* $\mathbf{C}^n$, J. Diif. Geom. **12** (1977), 499–511.

[**Ye**] Ye, Z., *On Nevanlinna's Erroe Terms,* preprint 1990.

[**Z**] Zaidenberg, M. G., *Stability of Hyperbolic Imbeddedness and Construction of Examples,* Math. USSR Sbornik **62** (1989), 331–361.

Some Recent Results and Problems in the Theory of Value-Distribution

Lo Yang

Dedicated to Professor Wilhelm Stoll on the occasion of his inauguration as the Duncan Professor of Mathematics.

For meromorphic functions of one complex variable, the theory of value-distribution has tremendously developed already since the twenties of this century. Although it has a long history, there are still some interesting and remarkable results during the recent years. For instance, Drasin [6] proved that the F. Nevanlinna conjecture is correct; Lewis and Wu [13] made a significant step in proving the Arakelyan's conjecture [1]; Osgood [18] and Steinmetz [20] independently proved the defect relation for small functions, and so on.

In this lecture, I would like to mention some recent results and problems which are based on my own interests.

1. Precise estimate of total deficiency of meromorphic functions and their derivatives

Let $f(z)$ be a transcendental meromorphic function in the finite plane and a be a complex value (finite or infinite). According to R. Nevanlinna

$$\begin{aligned}\delta(a,f) &= \liminf_{r\to\infty} \frac{m\left(r,\frac{1}{f-a}\right)}{T(r,f)} \\ &= 1-\limsup_{r\to\infty} \frac{N\left(r,\frac{1}{f-a}\right)}{T(r,f)}\end{aligned}$$

It is clear that $0 \leq \delta(a,f) \leq 1$. If $\delta(a,f)$ is positive, then a is named a deficient value with respect to $f(z)$ and $\delta(a,f)$ is its deficiency. The most fundamental result of Nevanlinna theory can be stated as follows [12, 17, 23].

Any transcendental meromorphic function $f(z)$ in the finite plane has countable deficient values at most and the total deficiency does not exceed 2. i.e.

$$\sum_{a\in\bar{\mathbb{C}}} \delta(a, f) \leq 2.$$

It is the famous Defect Relation (or Deficient Relation). In the general case, the upper bound 2 is sharp. If the order λ or the lower order μ of $f(z)$ is assigned, then the following deficient problem can be introduced. (Edrei [8])

Problem 1. Let $\mathcal{F}_\mu$, be the set of all the meromorphic functions of finite lower order μ. Can we determine

$$\Omega(\mu) = \sup_{f\in\mathcal{F}_\mu} \left\{ \sum_{a\in\bar{\mathbb{C}}} \delta(a, f) \right\}?$$

Do extremal functions exist? If so, what other properties characterize extremal functions?

When μ is less than 1, Edrei [8] obtained a precise estimate for the total deficiency by using the spread relation proved by Baernstein [2]. In fact, he proved

$$\Omega(\mu) = \begin{cases} 1, & 0 \leq \mu \leq \frac{1}{2}, \\ 2 - \sin \mu\pi, & \frac{1}{2} < \mu \leq 1. \end{cases}$$

The Problem 1, however, is still open for meromorphic functions of lower order bigger than 1, although a suitable bound has been suggested by Drasin and Weitsman [7] as follows:

$$\Omega(\mu) = \max\{\wedge_1(\mu), \wedge_2(\mu)\},$$

where

$$\wedge_1(\mu) = 2 - \frac{2 \sin \frac{\pi}{2}(2\mu - [2\mu])}{[2\mu] + 2 \sin \frac{\pi}{2}(2\mu - [2\mu])}$$

and

$$\wedge_2(\mu) = 2 - \frac{2\cos\frac{\pi}{2}(2\mu - [2\mu])}{[2\mu] + 1}.$$

Now we consider the derivative $f^{(k)}(z)$ of order k, where k is a positive integer. Hayman [11] pointed out that

$$\sum_{a\in\mathbb{C}} \delta(a, f^{(k)}) \leq \frac{k+2}{k+1}.$$

In 1971, Mues [15] improved this result to

$$\sum_{a\in\mathbb{C}} \delta(a, f^{(k)}) \leq \frac{k^2+5k+4}{k^2+4k+2}.$$

Recently I proved [26]

Theorem 1. Let $f(z)$ be a transcendental meromorphic function in the finite plane and k be a positive integer. Then we have

$$\sum_{a\in\mathbb{C}} \delta(a, f^{(k)}) \leq \frac{2k+2}{2k+1}.$$

It is clear that for any positive integer k, we always have

$$\frac{2k+2}{2k+1} < \frac{k^2+5k+4}{k^2+4k+2} < \frac{k+2}{k+1}$$

and

$$\frac{k^2+5k+4}{k^2+4k+2} - \frac{2k+2}{2k+1} > \frac{k+2}{k+1} - \frac{k^2+5k+4}{k^2+4k+2}.$$

Although Theorem 1 gives a much better estimate for $\sum_{a\in\mathbb{C}} \delta(a, f^{(k)})$, it does not include $\delta(\infty, f^{(k)})$. For this reason, we have another estimate [26].

Theorem 2. Let $f(z)$ be a transcendental meromorphic function of finite order in the finite plane and k be a positive integer. Then we have

$$\sum_{a\in\bar{\mathbb{C}}} \delta(a, f^{(k)}) \leq 2 - \frac{2k(1-\Theta(\infty, f))}{1+k(1-\Theta(\infty, f))},$$

where $\Theta(\infty, f)$ is the ramification index of ∞ with respect to f, defined by

$$\Theta(\infty, f) = 1 - \limsup_{r\to\infty} \frac{\overline{N}(r, f)}{T(r, f)}.$$

In particular, if $\Theta(\infty, f) < 1$, then we have

$$\lim_{k\to\infty} \left\{ \sum_{a\in\bar{\mathbb{C}}} \delta(a, f^{(k)}) \right\} = 0$$

It is natural to discuss the precise estimate of total deficiency of both the function itself and its derivative. For this subject, Drasin [5] posed the following questions.

Problem 2. Let $f(z)$ be meromorphic and of finite order in the finite plane. If $\sum_{a\in\overline{\mathbb{C}}} \delta(a, f) = 2$ and $\delta(\infty, f) = 0$, do we must have

$$\sum_{b\in\overline{\mathbb{C}}} \delta(b, f') = \delta(0, f') = 1?$$

Problem 3. Let $f(z)$ be meromorphic in the finite plane with $\delta(\infty, f) = 0$. Can we have

$$\sum_{a\in\bar{\mathbb{C}}} \delta(a, f) + \sum_{b\in\overline{\mathbb{C}}} \delta(b, f') = 4?$$

If not, what is best bound?

Quite recently, I proved the following theorem.

Theorem 3. Let $f(z)$ be a transcendental meromorphic function of finite order in the finite plane and k be a positive integer. Then we have

$$\sum_{a\in\mathbb{C}}\delta(a,f)+\sum_{b\in\overline{\mathbb{C}}}\delta(b,f^{(k)})\leq 3.$$

The equality holds if and only if either

(i) $\quad\Theta(\infty,f)=1,\quad \sum_{a\in\mathbb{C}}\delta(a,f)=1,\quad \sum_{b\in\overline{\mathbb{C}}}\delta(b,f^{(k)})=2;$

or

(ii) $k=1,\quad \Theta(\infty,f)=0,\quad \sum_{a\in\mathbb{C}}\delta(a,f)=2,\quad \sum_{b\in\overline{\mathbb{C}}}\delta(b,f')=1;$

Theorem 3 gives a positive answer to the Problem 2 by comparing Theorem 3 and the known fact

$$\sum_{a\in\mathbb{C}}\delta(a,f)\leq(2-\delta(\infty,f))\delta(o,f').$$

Theorem 3 answers also the Problem 3, when f is of finite order.

2. Conjectures of Frank, Goldberg and Mues

Mues [15] posed the following conjecture, when he improved the Hayman's estimate.

Problem 4. Let $f(z)$ be a transcendental meromorphic function in the finite plane and k be a positive integer. Then the following relation should be true.

$$\sum_{a\in\mathbb{C}}\delta(a,f^{(k)})\leq 1.$$

If $f(z)$ satisfies one of the following conditions, then the Mues conjecture can be easily verified.

(i) The order of f is finite and $\Theta(\infty,f)\leq 1-\frac{1}{k}$;
(ii) f has only poles with multiplicity $\leq k$;
(iii) The order of f is finite and

$$\delta(\infty,f^{(k)})+\sum_{a\in\mathbb{C}}\delta(a,f)\geq 2.$$

It seems that the Mues conjecture is true, when f has finite order.

Connecting the Problem 4, G. Frank and A. Goldberg recently raised two similar conjectures respectively.

Problem 5. Let $f(z)$ be a transcendental meromorphic function in the finite plane and k be a positive integer. If ε is an arbitrary small positive number, then the following estimate seems true.

$$k\overline{N}(r,f) < (1+\varepsilon)N(r,\frac{1}{f^{(k+1)}}) + S(r,f),$$

where

$$S(r,f) = O\{\log(rT(r,f))\},$$

except for r in a set with finite linear measure.

Problem 6. Let $f(z)$ be a transcendental meromorphic function in the finite plane. Then the following inequality should be correct.

$$\overline{N}(r,f) < N(r,\frac{1}{f''}) + S(r,f).$$

When $k \geq 2$, the Frank conjecture (Problem 5) is much stronger than the Goldberg conjecture (Problem 6). The Mues conjecture is a direct consequence of any one of them, when the order of $f(z)$ is finite.

3. Best coefficients of Hayman inequality

In 1959, Hayman [11] obtained a very interesting and remarkable inequality in which the characterstic function $T(r,f)$ can be bonnded by two counting functions. It is impossible without introducing the derivatives.

$$T(r,f) < (2+\frac{1}{k})N(r,\frac{1}{f}) + (2+\frac{2}{k})N(r,\frac{1}{f^{(k)}-1}) + S(r,f). \tag{3.1}$$

Later on, Hayman [12] adopted this inequality as the principal result of Chapter 3 in his book. He also posed the following question.

Problem 7. What are the best coefficients of the inequality (3.1).

Concerning this problem, the following inequality was obtained by the author [25] about two years ago.

Theorem 4. Let $f(z)$ be a transcendental meromorphic function in the finite plane and k be a positive integer. If ε is an arbitrary small positive number, then we have

$$T(r,f) < (1+\frac{1}{k}+\varepsilon)N(r,\frac{1}{f})+(1+\frac{1}{k}+\varepsilon)N(r,\frac{1}{f^{(k)}-1})+S(r,f).$$

The proof of Theorem 4 is based on the following lemma.

Lemma. Under the conditions of Theorem 4, we have

$$\overline{N}(r,f) < (\frac{1}{k}+\varepsilon)N(r,\frac{1}{f})+(\frac{1}{k}+\varepsilon)N(r,\frac{1}{f^{(k)}-1})+S(r,f).$$

If the Goldberg conjecture is true, then we have

$$T(r,f) < N(r,\frac{1}{f})+N(r,\frac{1}{f^{(k)}-1})+S(r,f). \qquad (3.2)$$

If the Frank conjecture is true, then we also have the inequality (3.2) in the case of $k \geq 2$ and

$$T(r,f) < (1+\varepsilon)N(r,\frac{1}{f})+(1+\varepsilon)N(r,\frac{1}{f'-1})+S(r,f)$$

in the ease of $k=1$.

4. Normal families and fix-points of meromorphic functions

A theorem which makes the connection between the normality of a given family of meromorphic functions and the lack of fix-points of both these functions and their derivatives, was proved by the author [24] in 1986.

Theorem 5. Let $\mathcal{F}$ be a family of meromorphic functions in a region D and k be a positive integer. If, for every function $f(z)$ of $\mathcal{F}$, both $f(z)$ and $f^{(k)}(z)$ (the derivative of order k) have no fix-points in D, then $\mathcal{F}$ is normal there.

Since the iteration of $f(z)$ is very important and grows much faster than $f(z)$ and $f^{(k)}(z)$, it is natural to pose the following problem.

Problem 8. Let $\mathcal{F}$ be a family of entire functions, D be a region and k be a fixed positive integer. If, for every function $f(z)$ of $\mathcal{F}$, both $f(z)$ and $f^k(z)$ (the iteration of order k of $f(z)$) have no fix-points in D, is $\mathcal{F}$ normal there?

Schwick [19] proved several criteria for normality. Among others, he proved

Theorem 6. Let $\mathcal{F}$ be a family of meromorphic functions in a region D and n and k be two positive integers. If for every function $f(z)$ of $\mathcal{F}$,

$$(f^n)^{(k)} \neq 1$$

and

$$n \geq k + 3, \tag{4.1}$$

then $\mathcal{F}$ is normal in D. Moreover, if $\mathcal{F}$ is a family of holomorphic functions, then the condition (4.1) can be replaced as

$$n \geq k + 1. \tag{4.2}$$

It seems to me that the following assertion should be true.

Problem 9. Let $\mathcal{F}$ be a family of meromorphic functions in a region D and n and k be two positive integers with (4.1). If, for every function $f(z)$ of $\mathcal{F}$, $(f^n)^{(k)}$ has no fix-points, then $\mathcal{F}$ is normal in D. When $\mathcal{F}$ is a family of holomorphic functions, then (4.1) can also be replaced by (4.2).

5. Common Borel directions of a meromorphic function and its derivatives

Let $f(z)$ be meromorphic and of order λ in the finite plane, where $0 < \lambda < \infty$. Valiron [21] proved there exists a direction B : $\arg z = \theta_0 (0 \leq \theta_0 < 2\pi)$ such that, for any positive number ε and any complex value a, we always have

$$\limsup_{r \to \infty} \frac{\log n(r, \theta_0, \varepsilon, f = a)}{\log r} = \lambda,$$

except two values of a at most, where $n(r, \theta_0, \varepsilon, f = a)$ denotes the number of zeros of $f - a$ in the region $(|z| \leq r) \cap (|\arg z - \theta_0| \leq \varepsilon)$. Such direction is named a Borel direction of order λ of $f(z)$.

Since $f^{(k)}(z)$ is also meromorphic and of order λ in the finite plane, it has a Borel direction too. Valiron [21] asked the following question.

Problem 10. Let $f(z)$ be meromorphic and of order λ in the finite plane, where $0 < \lambda < \infty$. Is there a common Borel direction of $f(z)$ and all its derivatives?

Concerning this problem, the known result is

Theorem 7. Let $f(z)$ be meromorphic and of order λ in the finite plane, where $0 < \lambda < \infty$. If $f(z)$ has a Borel exceptional value (which is either a finite complex value or the infinity), then there exists a common Borel direction of $f(z)$ and all its derivatives.

The papers concerning Theorem 7 are due to Milloux, Zhang K. H., Zhang Q. D. and myself [14, 23, 27].

On the other hand, the following fact is also known.

There exists a meromorphic function $f_0(z)$ of order one in the finite plane such that its derivative $f_0'(z)$ has more Borel directions than $f_0(z)$.

For instance, the function

$$f_0(z) = \frac{e^{-iz}}{1 + e^z},$$

which was pointed out by Steinmetz, is such an example. $f_0(z)$ has the Borel directions $\arg z = \frac{\pi}{4}$, π and $\frac{3\pi}{2}$, but $f_0'(z)$ has these and in addition the direction $\arg z = \frac{\pi}{2}$.

6. Optimum condition to ensure the existence of Hayman direction

Corresponding to the Hayman's inequality, we may ask if there are some similar results in the angular distribution. My following theorem [22] aims at answer of this question.

Theorem 8. Let $f(z)$ be a meromorphic function in the finite plane. If

$$\limsup_{r\to\infty} \frac{T(r,f)}{(\log r)^3} = \infty, \tag{6.1}$$

then there exists a direction (H) : $\arg z = \theta_0$ such that, for any positive number ε, an arbitrary integer k and any two finite complex values a and $b (b \neq 0)$, we have

$$\lim_{r\to\infty} \{n(r,\theta_0,\varepsilon, f=a) + n(r,\theta_0,\varepsilon, f^{(k)}=b)\} = \infty.$$

It is convenient to name such kind of direction as Hayman direction.

For a meromorphic function, since the condition ensuring a Julia direction is

$$\limsup_{r\to\infty} \frac{T(r,f)}{(\log r)^2} = \infty, \tag{6.2}$$

Drasin [3] raised the following question in 1984.

Problem 11. Is Theorem 8 still true, if the condition (6.1) is replaced by (6.2)?

It seems that the answer of Problem 11 is positive, since Chen H. H. [4] proved recently the following fact.

Theorem 9. Let $f(z)$ be a meromorphic function in the finite plane. If the condition (6.2) is satisfied, then there exists a direction

(H) : arg $z = \theta_0$ such that for any positive number ε, an arbitrary positive integer k and any two finite complex values a and $b (b \neq 0)$, we have

$$\limsup_{r\to\infty} \frac{n(r, \theta_0, \varepsilon, f = a) + n(r, \theta_0, \varepsilon, f^{(k)} = b)}{\log r} = \infty.$$

7. Picard type theorems and the existence of singular directions

Bloch principle says a family of holomorphic (or meromorphic) functions in a region satisfying a condition (or a set of conditions) uniformly which can only be possessed by the constant functions in the finite plane, must be normal there. In simple words, there usually is a criterion for normality to correspond a Picard type theorem. Similarly we have the following question.

Problem 12. Corresponding to every Picard type theorem, is there a singular direction? To be precise, let P be a property (or a set of properties) such that any entire function (or a meromorphic function in the finite plane) satisfying P, must be a constant. Then for any transcendental entire function (or a meromorphic function with some suitable condition of growth), is there a ray arg $z = \theta_0 (0 \leq \theta_o < 2\pi)$ such that $f(z)$ does not satisfy P in the angle $|\arg z - \theta_0| < \varepsilon$, for any small positive number ε.

The direction in the Problem 12 is a direction of Julia type. We can also pose a problem for a direction of Borel type.

For instance, the following fact is well known.

Let $f(z)$ be meromorphic in the finite plane. If $(f^n)^{(k)} \neq 1$ for two positive integers n and k with $n \geq k+3$, then $f(z)$ must reduce to a constant. When $f(z)$ is entire, the condition $n \geq k+1$ is sufficient.

Problem 13. Let $f(z)$ be meromorphic and of order λ in the finite plane, where $0 < \lambda < \infty$. Is there a direction arg $z = \theta_0 (0 \leq \theta_0 < 2\pi)$ such that for any positive number ε two arbitrary positive integers n and k with $n \geq k + 3$ and any finite, non-zero complex value a, we have

$$\limsup_{r\to\infty} \frac{\log n(r, \theta_0, \varepsilon, (f^n)^{(k)} = a)}{\log r} = \lambda?$$

When $f(z)$ is entire, the condition $n \geq k+1$ seems sufficient.

8. Growth, number of deficient values and Picard type theorem

Picard type theorems are not only involved in criteria for normality and singular directions, but also connect with the growth and number of deficient values.

Problem 14. Let P be a property (or a set of properties) given by the Problem 12. Suppose that $f(z)$ is entire (or meromorphic) and of finite lower order μ in the finite plane and that $L_j : \arg z = \theta_j (j = 1, 2, \cdots, J; 0 \leq \theta_1 < \theta_2 < \cdots < \theta_J < 2\pi)$ are finite number of rays issued from the origin. If $f(z)$ satisfies P in $\mathbb{C} \backslash (\cup_{j=1}^{J} L_j)$ then the order λ of $f(z)$ seems to have the estimate

$$\lambda \leq \max_{1 \leq j \leq J} \left\{ \frac{\pi}{\theta_{j+1} - \theta_j} \right\}, \quad (\theta_{J+1} = \theta_1 + 2\pi)$$

and the number of finite non-zero deficient values does not exceed J.

The property P can be chosen as

(i) $f(z) \neq 0$ and $f^{(k)}(z) \neq 1 (k \in \mathbb{Z}^+)$;

(ii) $f'(z) - af(z)^n \neq b$, where a and b are two finite complex values with $a \neq 0$ and $n \geq 5$ is a positive integer;

(iii) $(f(z)^n)^{(k)} \neq 1$, where n and k are two positive integers with $n \geq k+3$, when f is meromorphic and $n \geq k+1$, when f is entire.

9. Value-distribution with respect to small functions

Let $f(z)$ be meromorphic in the finite plane and a be a complex value. The theory of value-distribntion investigates the distribution of zeros of $f(z) - a$. It is natural to instead of the complex value a by another meromorphic function $a(z)$ with the condition

$$T(r, a(z)) = o\{T(r, f)\}. \tag{9.1}$$

Therefore, corresponding to results in the theory of value-distribution, we can ask similar questions with respect to small functions. Some of them are very difficult and significant. For instance, R. Nevanlinna himself asked if his defect relation can be extended to small functions. Up to few years ago, Osgood [18] and Steinmetz [20] independently settled this problem with a positive answer.

Theorem 10. Let $f(z)$ be a transcendental meromorphic function in the finite plane and A be the set of all meromorphic functions $a(z)$ with the condition (9.1). Then we have

$$\sum_{a(z)\in A} \delta(a(z), f) \leq 2.$$

All the complex values including the infinity are contained in A.

There are still some problem. We mention only one of them here.

Recently, Lewis and Wu [13] proved

Theorem 11. If $f(z)$ is an entire function of finite lower order, then there exists a positive number τ_0 not depending on f such that the series $\sum_{a\in\overline{\mathbb{C}}} (\delta(a, f))^{\frac{1}{3}-\tau}$ is convergent for any τ with $0 < \tau < \tau_0$.

Problem 15. Is Lewis and Wu's result still true for small functions?

Acknowledgement. The author is very grateful to the Department of Mathematics, University of Notre Dame, Prof. W. Stoll and Prof. P. M. Wong for their hospitality.

REFERENCES

[1] Arakelyan, N. U., *Entire function of finite order with a set of infinite deficient values (in Russian).* Dokal. USSR, 170 (1966), 999–1002.

[2] Baernstein, A., *Proof of Edrei's spread conjecture*, Proc. London Math. Soc., 26 (1973), 418–434.

[3] Barth, K. F., Brannan, D. A. and Hayman, W. K., *Research Problems in Complex Analysis*, Bull. London Math. Soc., 16 (1984), 490–517.

[4] Chen, H. H., *Singular directions corresponding to Hayman's inequality* (Chinese), Adv. in Math. (Beijing), 16 (1987), 73–80.

[5] Drasin, D., *An introduction to potential theory and meromorphic functions, Complex analysis and its applications*, Vol. 1, IAEA, Vienna, 1976, 1–93.

[6] Drasin, D., *Proof of a conjecture of F. Nevanlinna concerning functions which have deficiency sum two*, Acta Math., 158 (1987), 1–94.

[7] Drasin, D. and Weitsman, A., *Meromorphic functions with large sums of deficiencies*, Adv. in Math., 15 (1974), 93–126.

[8] Edrei, A., *Solution of the deficiency problem for functions of small lower order*, Proc. London Math. Soc., 26 (1973), 435–445.

[9] Edrei, A. and Fuchs, W. H. J., *On meromorphic functions with regions free of poles and zeros*, Acta Math., 108 (1962), 113–145.

[10] Frank, G. and Weissenborn, G., *Rational deficient functions of meromorphic functions*, Bull. London Math. Soc., 18 (1986), 29-33.

[11] Hayman, W. K., *Picard values of meromorphic functions and their derivatives*, Ann. of Math., 70 (1959), 9–42.

[12] Hayman, W. K., *Meromorphic functions*, Oxford, 1964.

[13] Lewis, J. L. and Wu, J. M., *On conjectures of Arakelyan and Littlewood*, J. d'Analyse Math., 50 (1988), 259–283.

[14] Milloux, H., *Sur les directions de Borel des fonctions entiére, de leurs derivées et de leurs integrales*, J. d'Analyse Math., 1 (1951), 244–330.

[15] Mues, E., *Über eine Defekt und Verzweigungsrelation für die Ableitung Meromorpher Funktionen*, Manuscripta Math., 5 (1971), 275–297.

[16] Nevanlinna, R., *Le Théorème de Picard-Borel et la théorie des fonctions méromorphes*, Coll Borel, 1929.

[17] Nevanlinna, R., *Analytic functions*, Springer-Verlag, Berlin, 1970.

[18] Osgood, C. F., *Sometimes effective Thue-Siegel-Schmidt-Nevanlinna bounds or better*, J. Number theory, 21 (1985), 347–389.

[19] Schwick, W., *Normality Criteria for families of meromorphic functions*, J. D'Analyse Math., 52 (1989), 241–289.

[20] Steinmetz, N., *Eine Verallgemeinerung des zweiten Nevanlinnaschen Hauptsatzes*, J. für Math., 368 (1986), 134–141.

[21] Valiron, G., *Recherches sur le théorème de M. Borel dans la théorie des fonctions méromorphes*, Acta Math., 52 (1928), 67–92.

[22] Yang, Lo, *Meromorphic functions and their derivatives*, J. London Math. Soc., (2) 25 (1982), 288–296.

[23] Yang, Lo, *Theory of value-distribution and its new research* (in Chinese), Science Press, Beijing, 1982.

[24] Yang, Lo, *Normal families and fix-points of meromorphic functions*, Indiana Univ. Math. J., 35 (1986), 179–191.

[25] Yang, Lo, *Precise fundamental inequalities and sum of deficiencies*, Sci. Sinica, **34** (1991), 157–165.

[26] Yang, Lo, *Precise estimate of total deficiency of meromorphic derivatives*, J. d'Analyse Math., **55** (1990), 287–296.

[27] Yang, Lo and Zhang Qingde, *New singular directions of meromorphic functions*, Sci. Sinica, 27 (1984), 352-366.